I0427479

DIVE INTO COUNTING WITH FINLEY FISH

This book belongs to:

_____

_____

ADVENTURES IN SKILLVILLE

Join Finley Fish, Olivia Owlington, Ollie the Octopus, Harmony Hedgehog, and Sammy Snail in the enchanting town of Skillville, where learning is always an adventure!

Each character specializes in a unique set of early developmental skills, creating a diverse and engaging series that caters to children's holistic growth. and skills mastery.

EMAIL US AT
MextraPublishing@gmail.com
To get FREE Bonus downloadable worksheets and previews of upcoming books!
Title the email "Skillville Bonus"

MEET FINLEY FISH

Embark on a fin-tastic mathematical journey with Finley Fish as your child's underwater guide in the enchanting realm of Skillville Math.

Finley Fish specializes in introducing young learners to the world of numbers. From counting colorful sea creatures to recognizing and understanding basic math skills, Finley's adventures make numerical concepts approachable and enjoyable.

Always eager to dive into the excitement of learning, Finley transforms every adventure into a math-filled discovery. Children will join Finley in counting seashells, grouping seaweed in sets, and navigating coral reefs, all while practicing number recognition.

Finley's tail flips with excitement as he guides children through the enchanting ocean of mathematics, fostering a love for learning and a solid foundation in basic math skills that will serve them swimmingly in their educational journey.

NUMBER TRACING

NUMBER TRACING

ONE

1

1 1 1 1 1

1 1 1 1 1

1 1 1 1 1 1 1 1

one

COUNT AND COLOR

FIND AND COLOR EVERY NUMBER ONE

1	8	5	3	9
7	4	2	1	6
5	1	6	8	1

COLOR 1 SEAWEED:

TWO

2

2 2 2
2 2 2
2 2 2 2 2

two

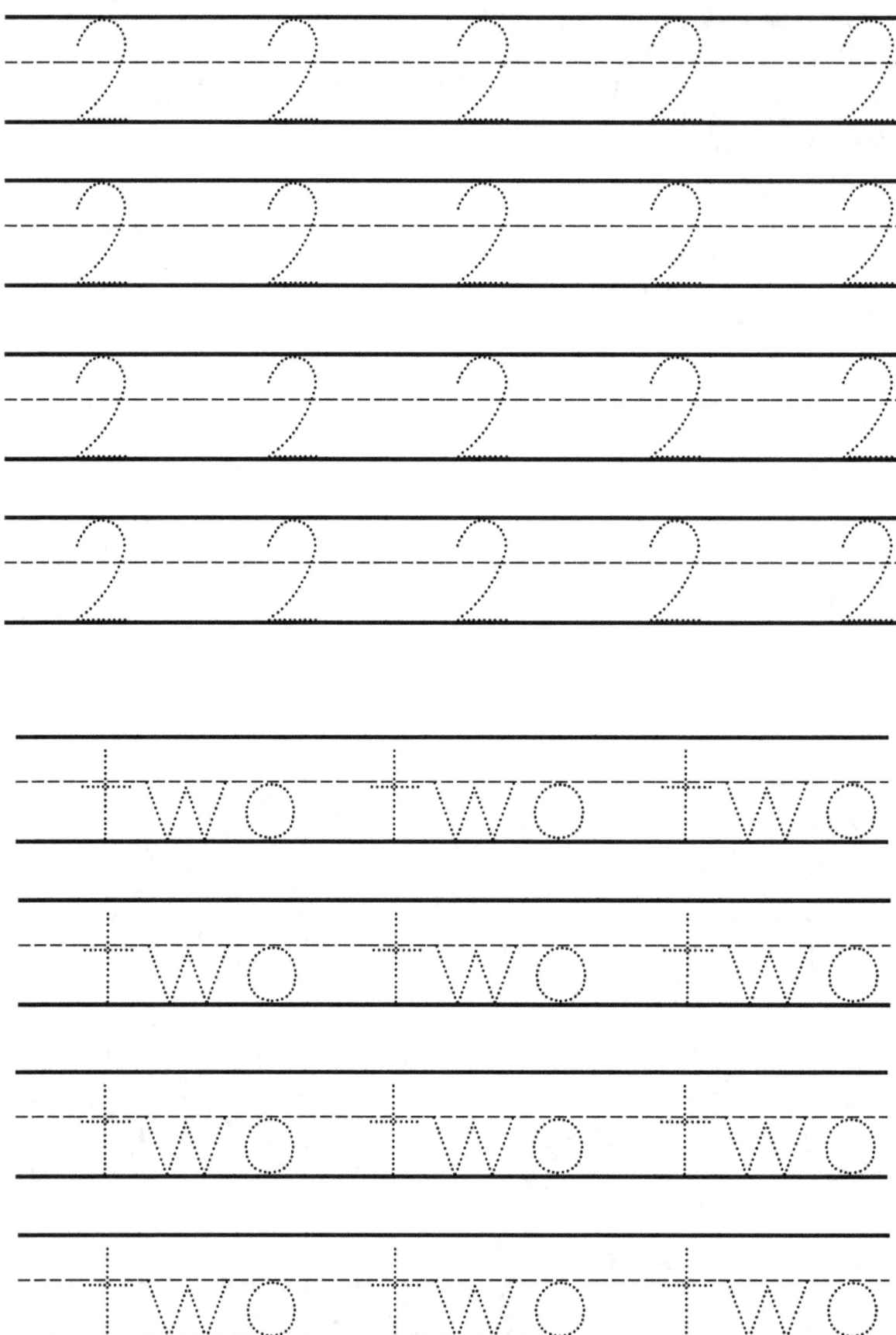

COUNT AND COLOR

FIND AND COLOR EVERY NUMBER TWO

9 7 2 4 1

5 2 3 8 2

2 5 9 2 6

COLOR 2 WHALES:

THREE

3

three

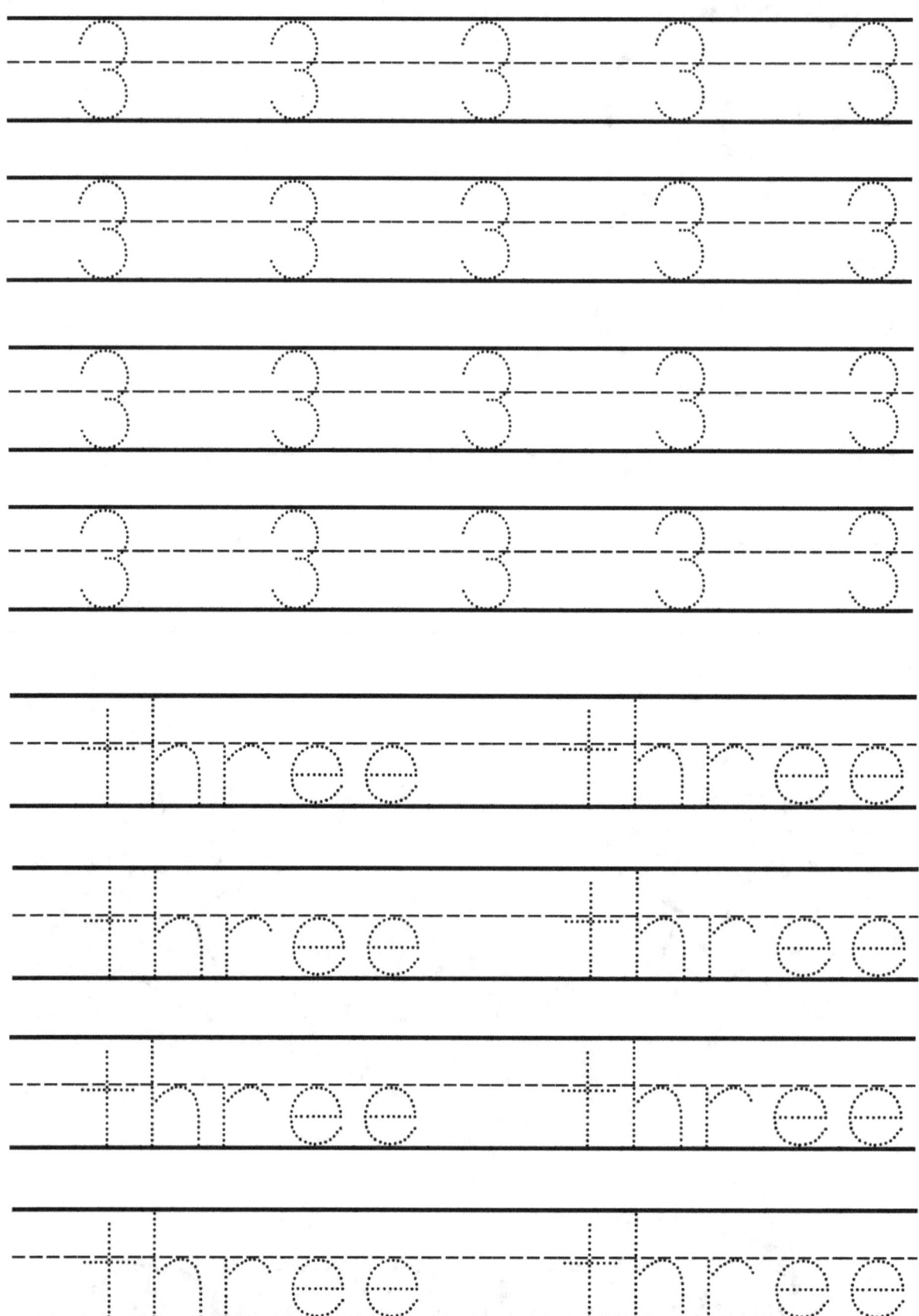

COUNT AND COLOR

FIND AND COLOR EVERY NUMBER THREE

3	8	6	3	7
5	3	4	8	9
1	6	3	2	8

COLOR 3 JELLY FISH:

FOUR

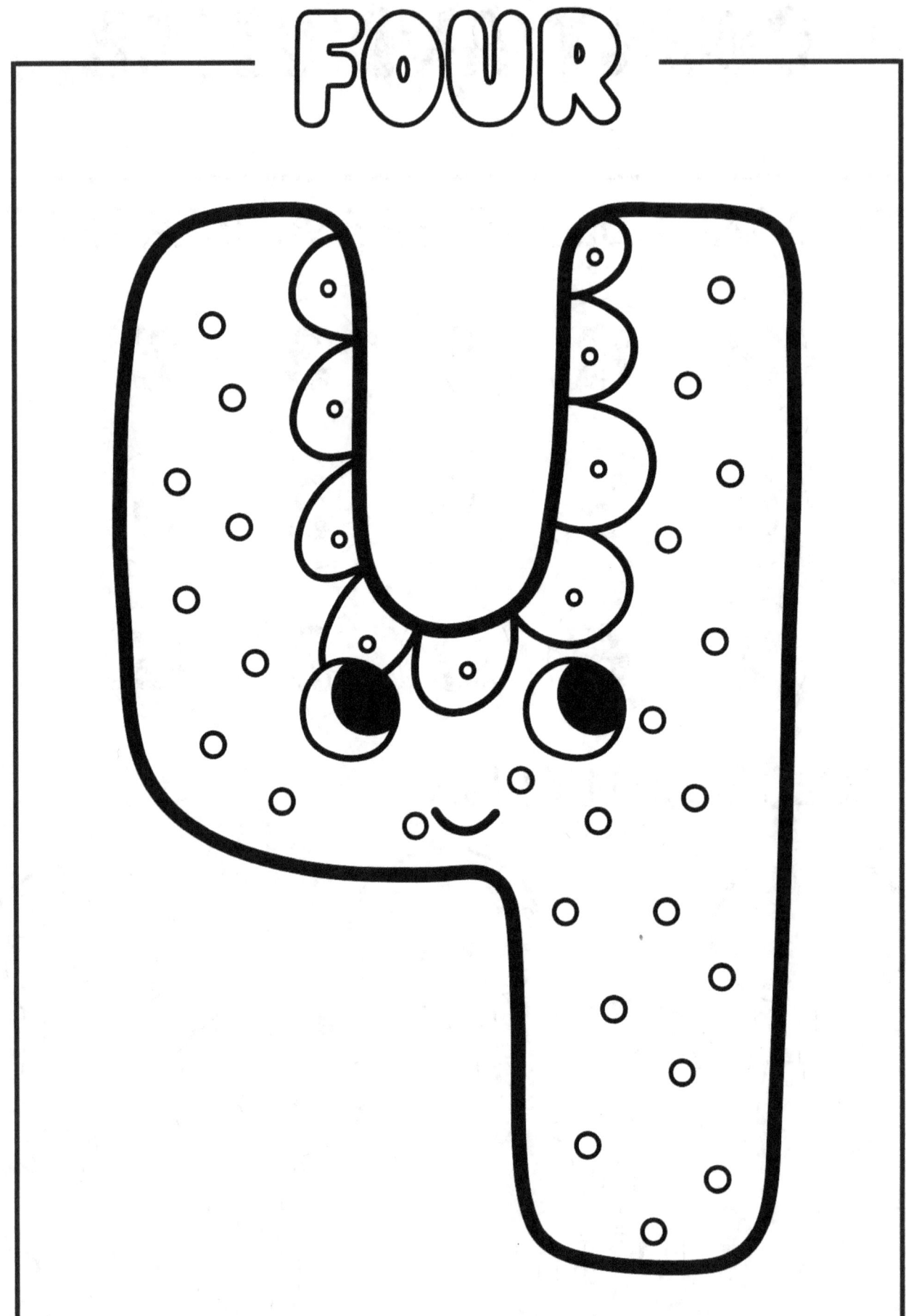

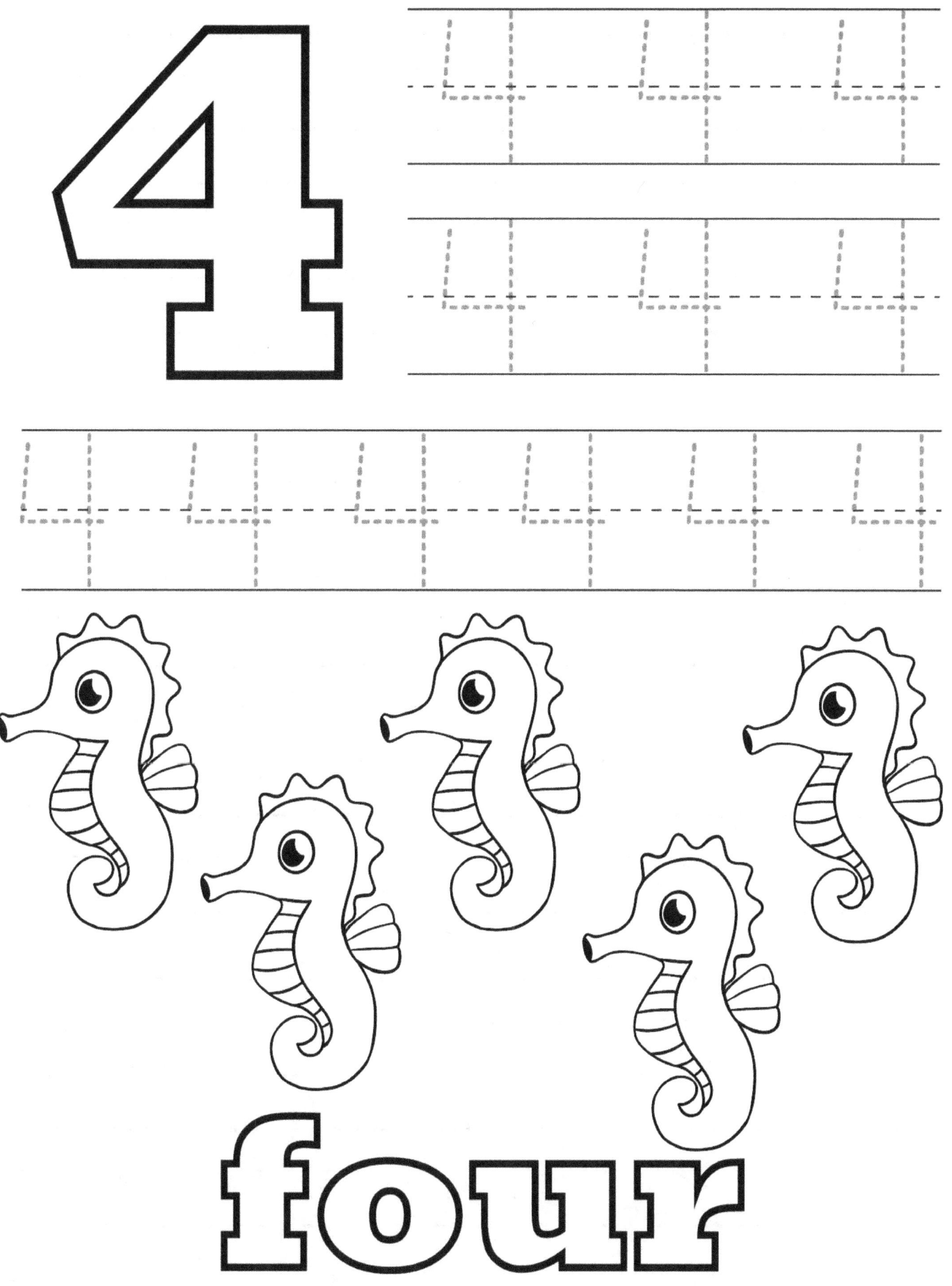

four

four four four
four four four
four four four
four four four

COUNT AND COLOR

FIND AND COLOR EVERY NUMBER FOUR

1	5	4	8	6
2	4	9	7	2
3	1	4	2	4

COLOR 4 CRABS:

FIVE

5

5 5 5

5 5 5

5 5 5 5 5

five

COUNT AND COLOR

FIND AND COLOR EVERY NUMBER FIVE

5	8	4	9	5
2	5	7	5	3
1	6	5	2	5

COLOR 5 TURTLES:

SIX

six

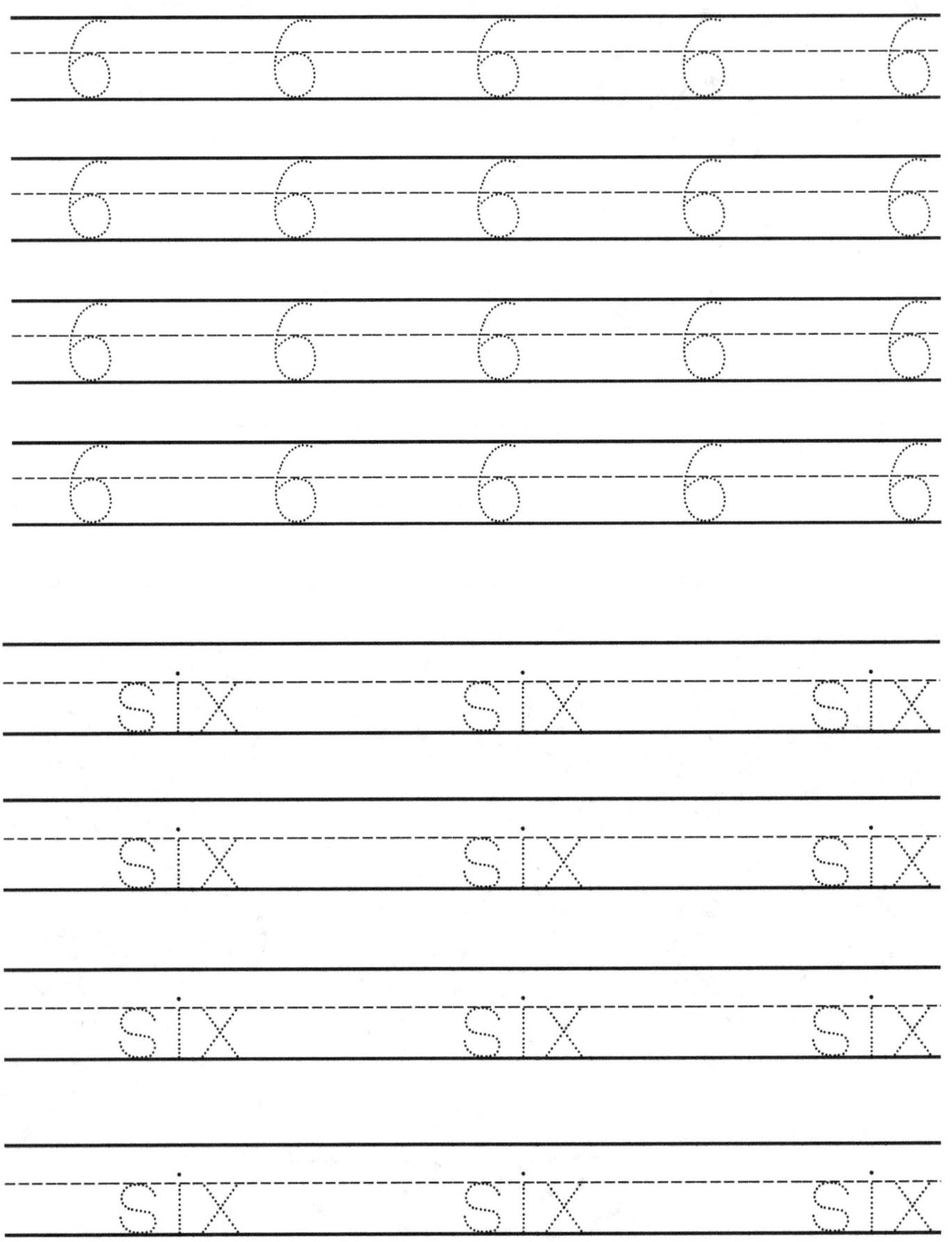

COUNT AND COLOR

FIND AND COLOR EVERY NUMBER SIX

8	6	4	7	9
1	9	6	3	6
6	4	8	6	5

COLOR 6 DOLPHINS:

SEVEN

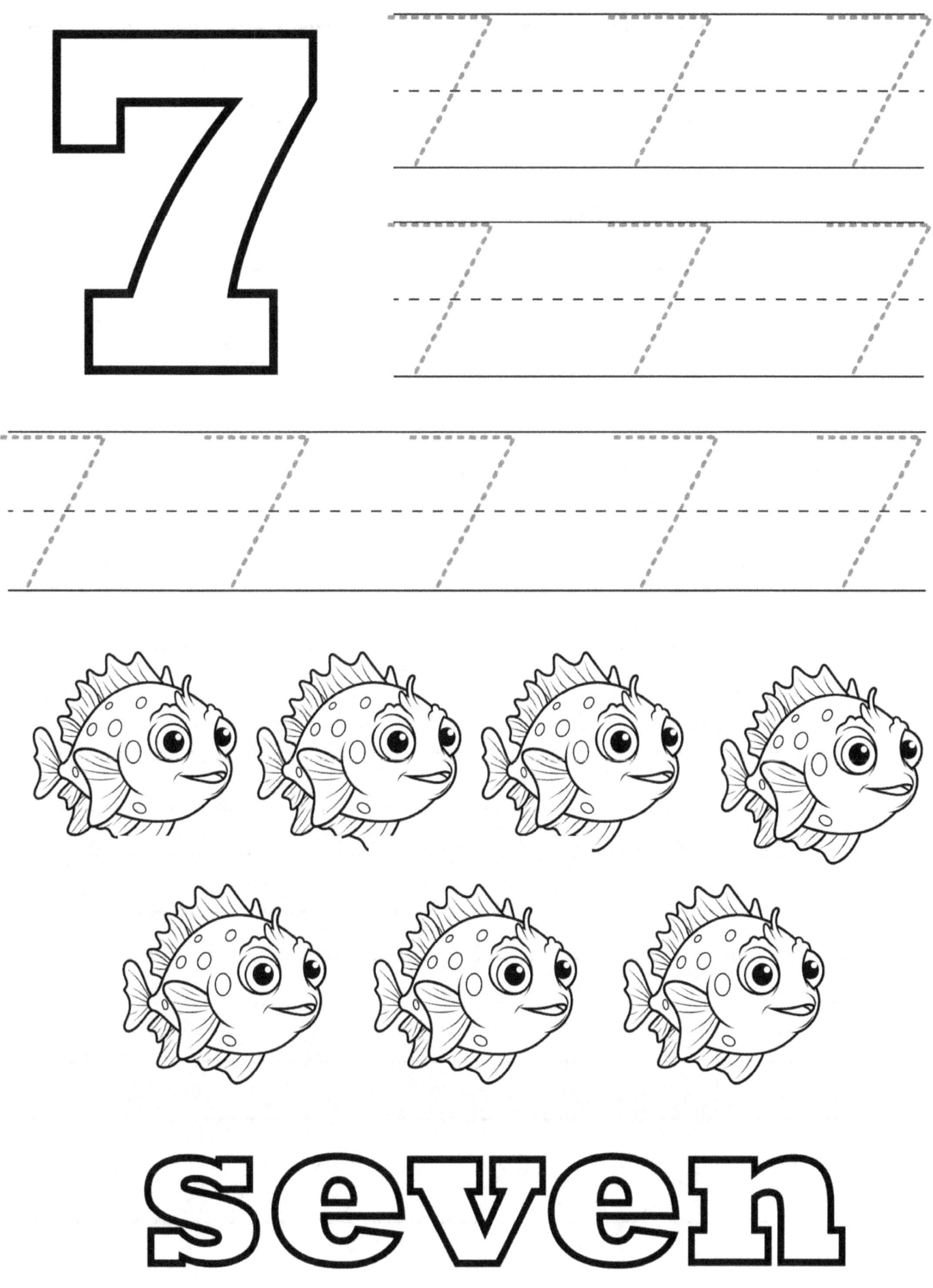

7

seven

COUNT AND COLOR

FIND AND COLOR EVERY NUMBER SEVEN

8 6 4 7 9

1 9 6 3 6

6 4 8 6 5

COLOR 7 SHARKS:

EIGHT

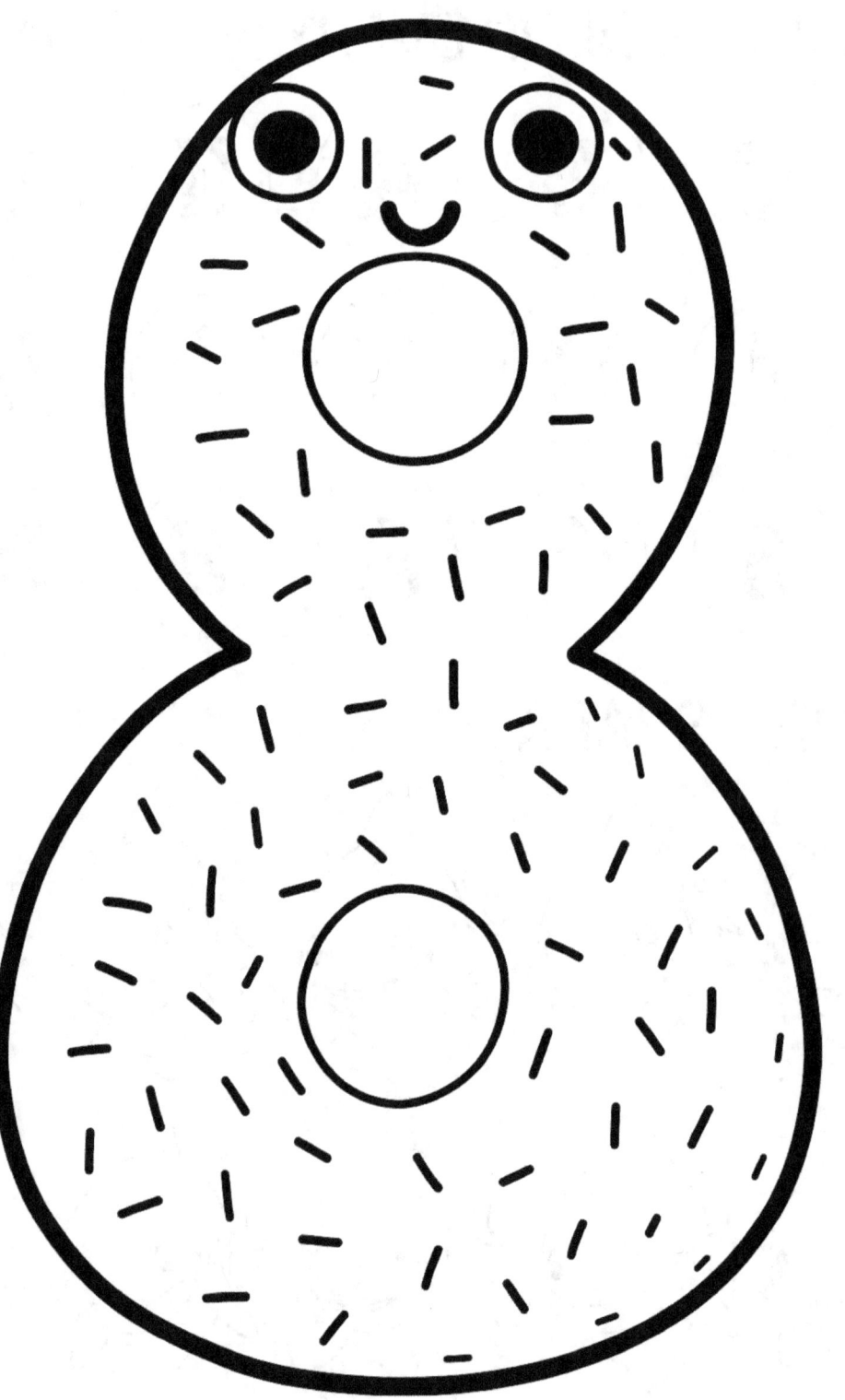

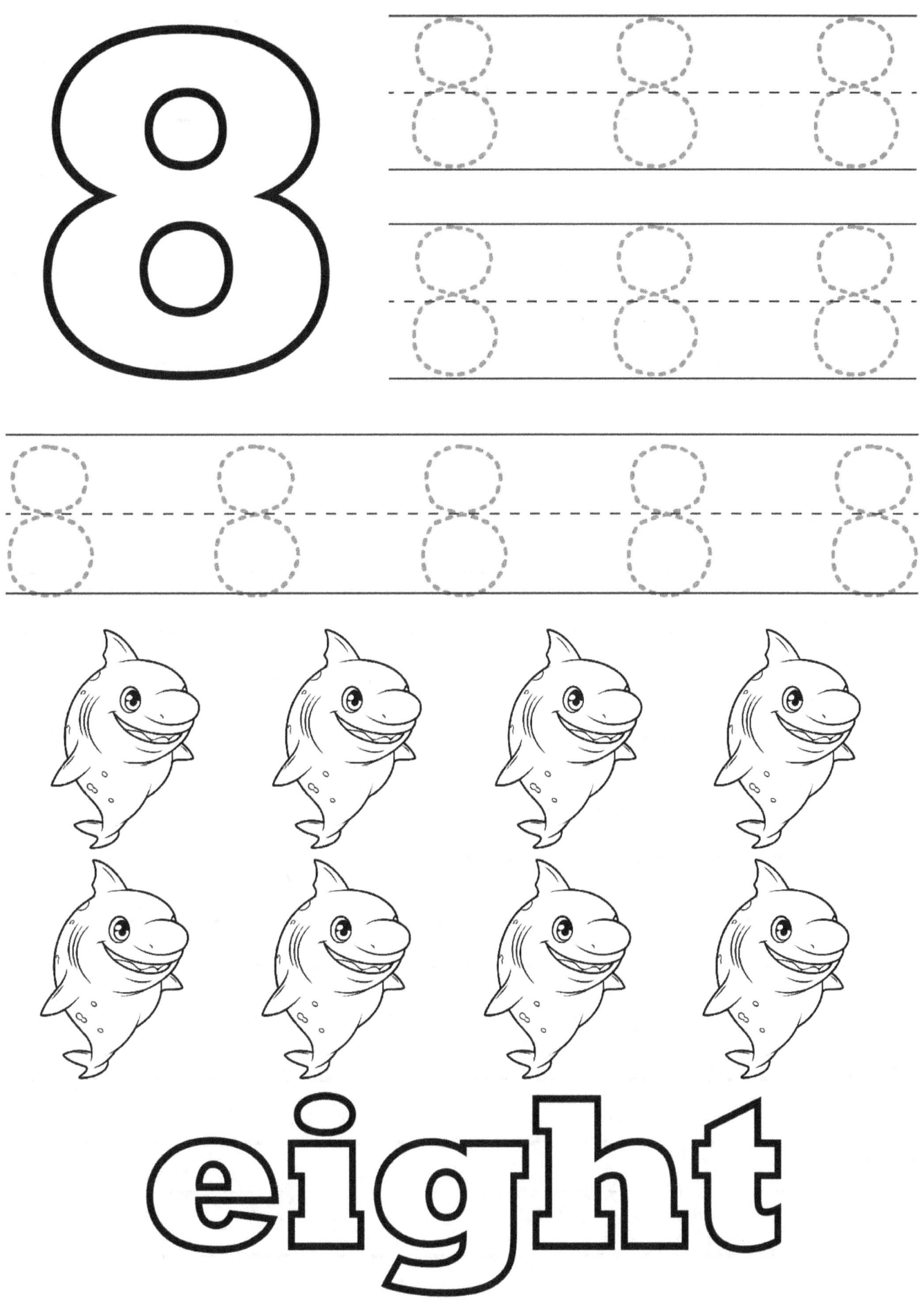

eight

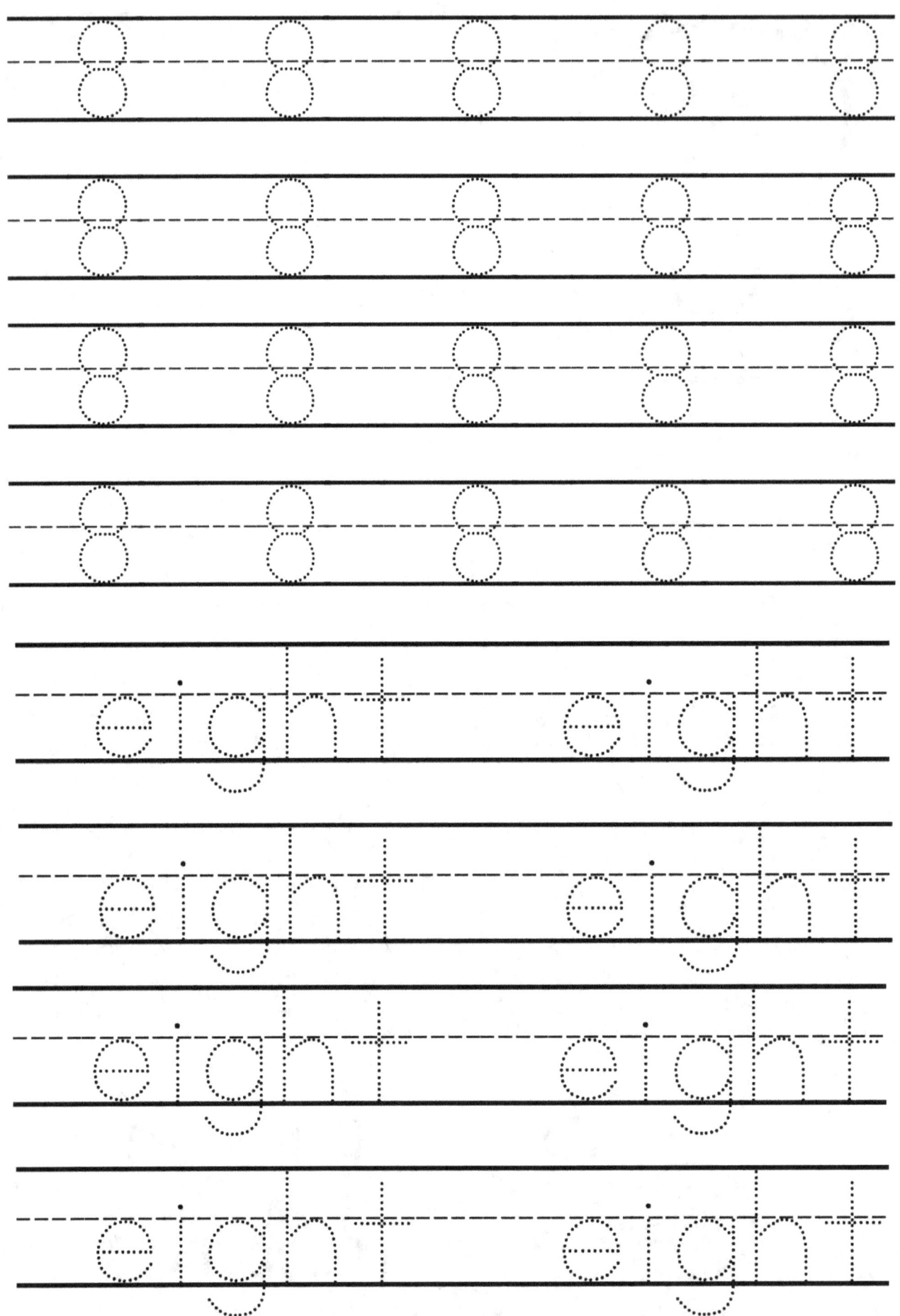

COUNT AND COLOR

FIND AND COLOR EVERY NUMBER EIGHT

5	8	3	8	2
1	7	8	4	8
3	6	9	6	8

COLOR 8 TURTLES :

NINE

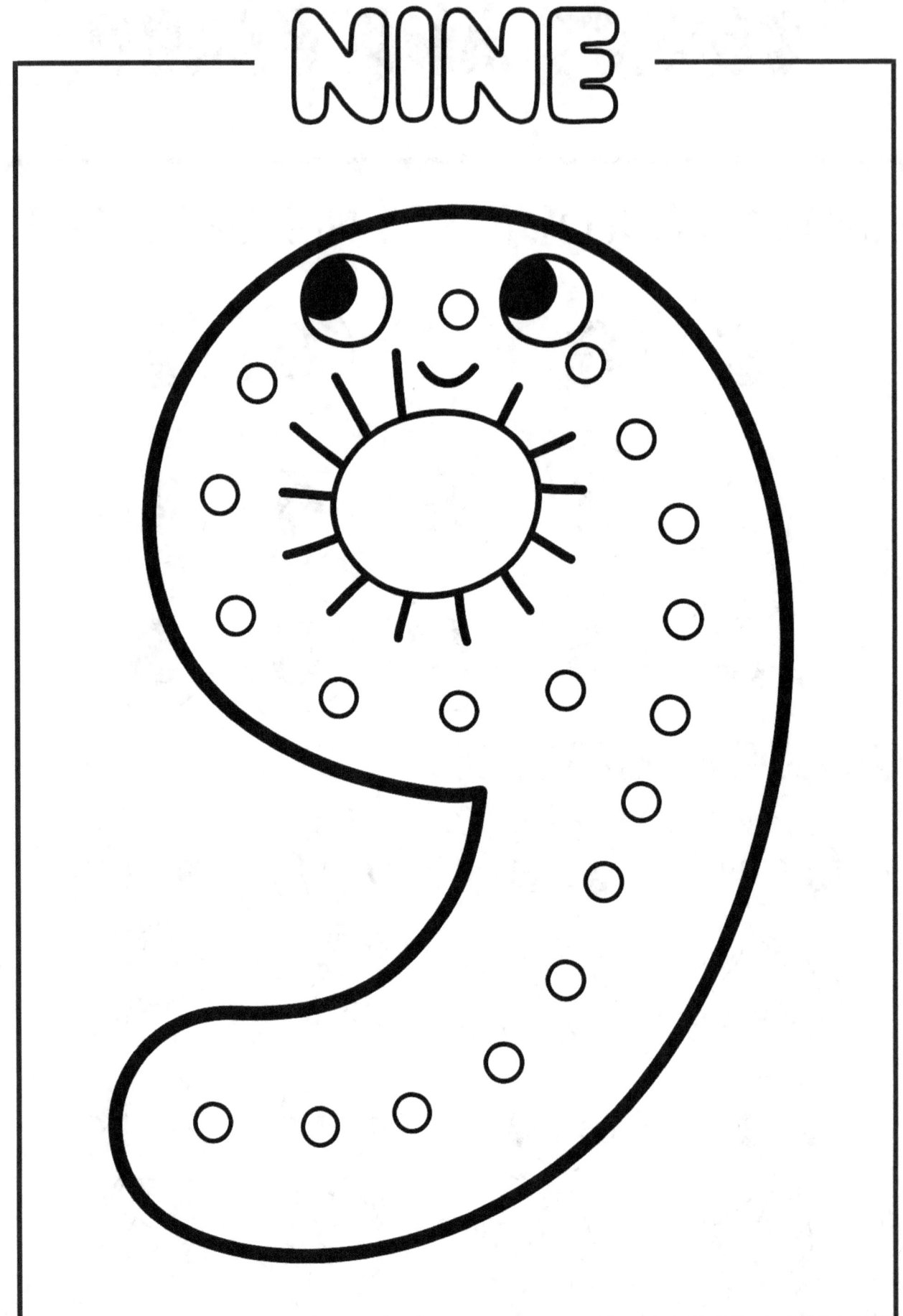

nine

COUNT AND COLOR

FIND AND COLOR EVERY NUMBER NINE

9	8	3	9	2
1	7	9	4	9
9	6	5	9	8

COLOR 9 STARFISH:

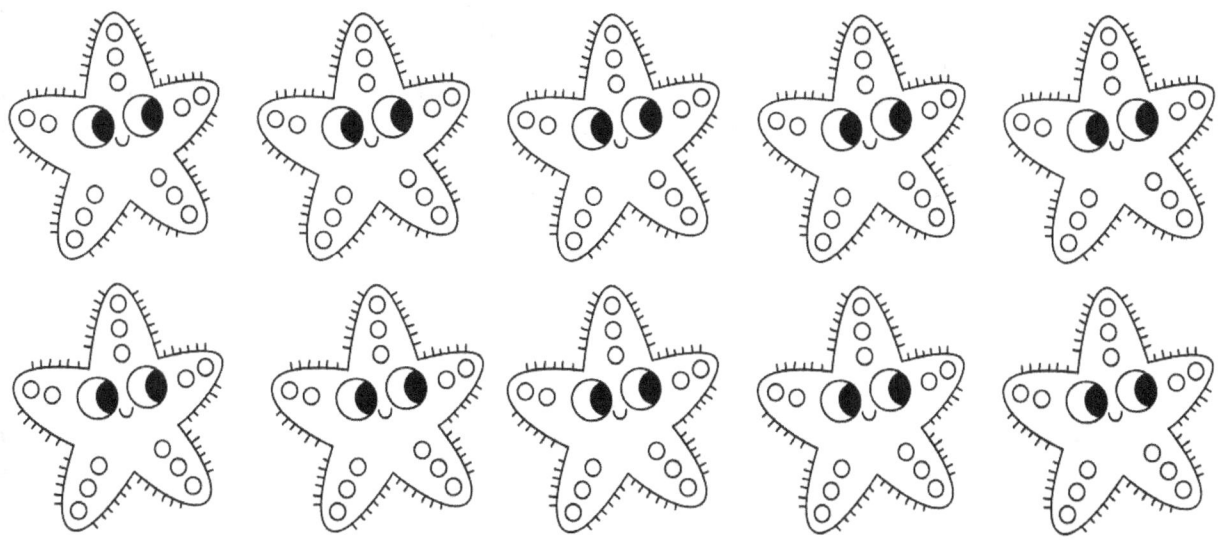

TEN

10

10 10

10 10

10 10 10 10 10

ten

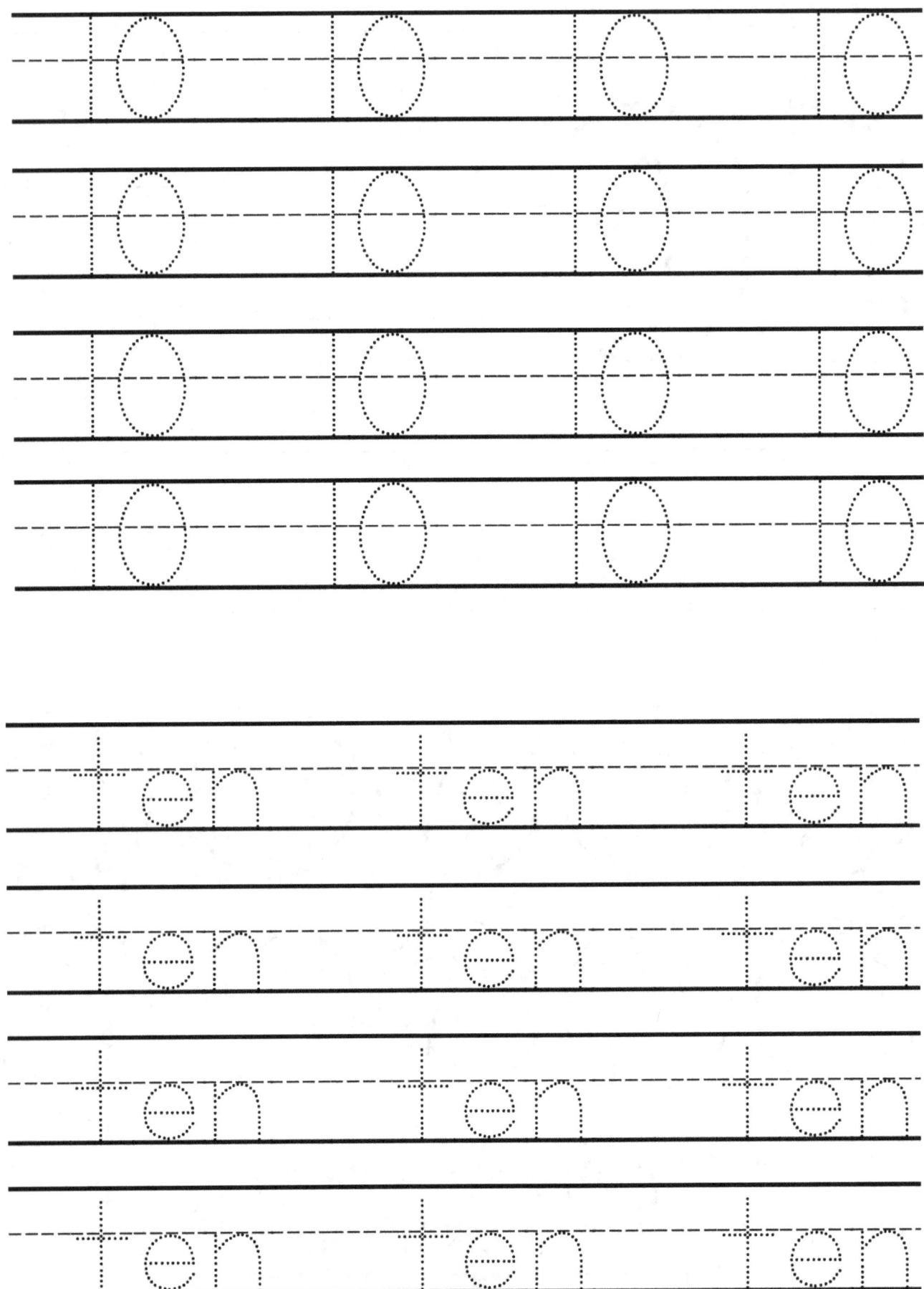

COUNT AND COLOR

FIND AND COLOR EVERY NUMBER TEN

3 7 10 5 10

10 2 1 10 6

8 10 4 9 10

COLOR 10 EELS

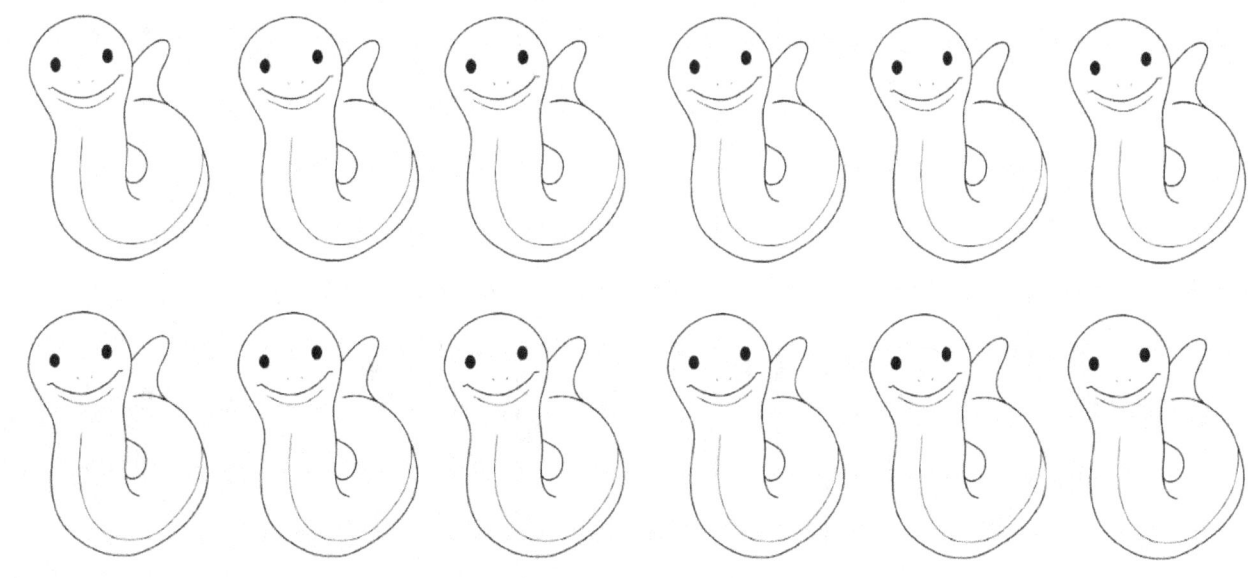

NUMBER TRACING

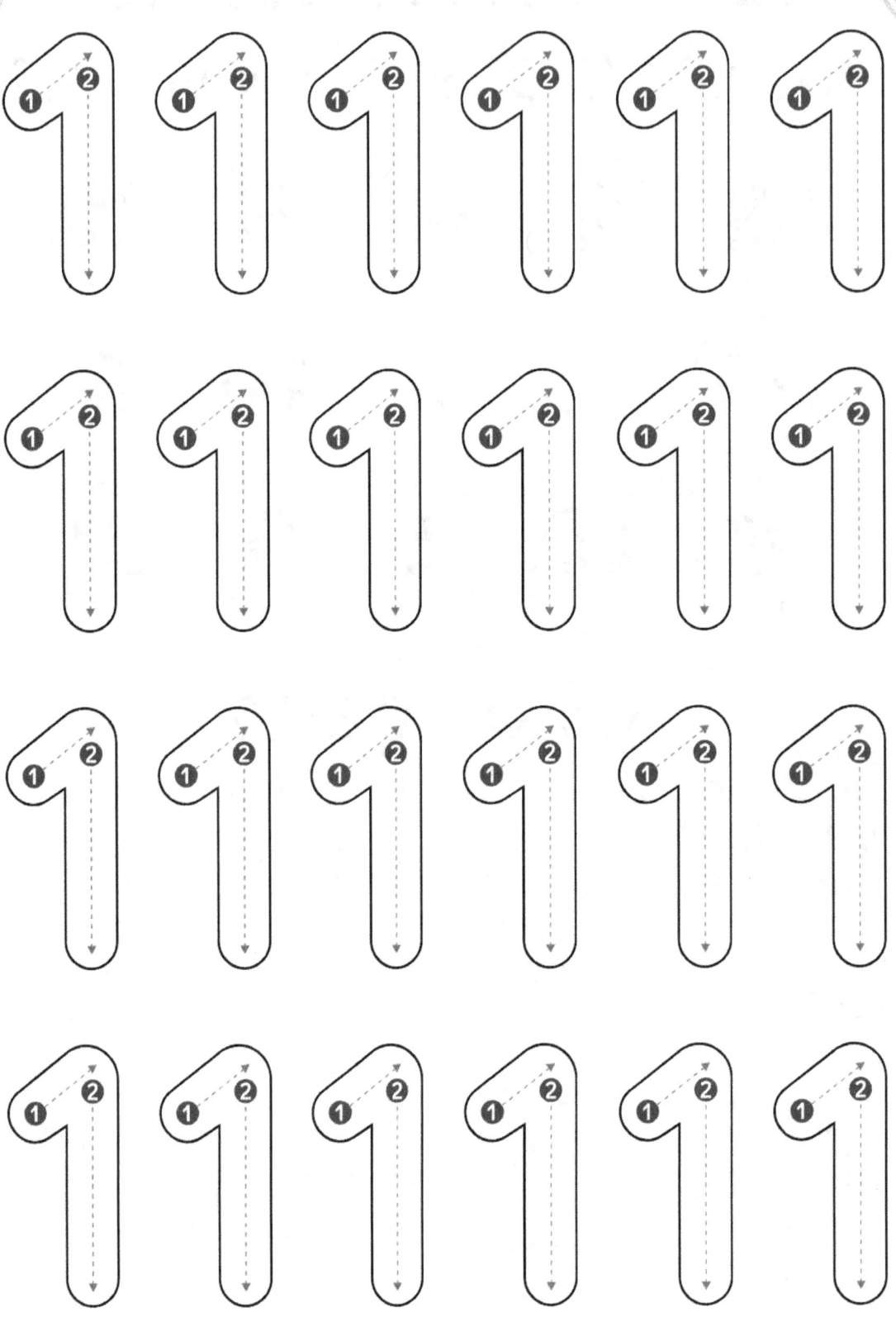

NUMBER TRACING

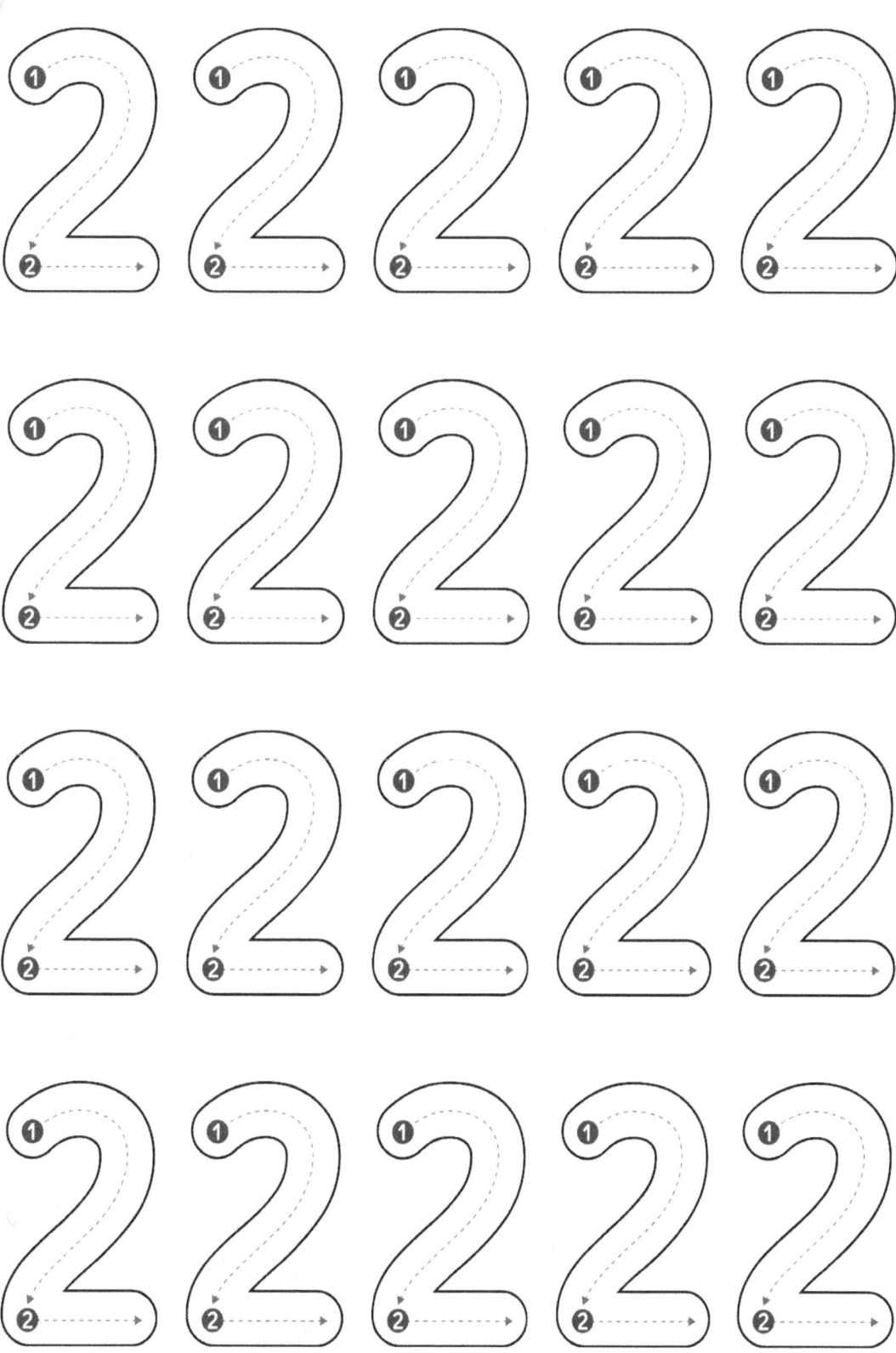

NUMBER TRACING

NUMBER TRACING

NUMBER TRACING

NUMBER TRACING

NUMBER TRACING

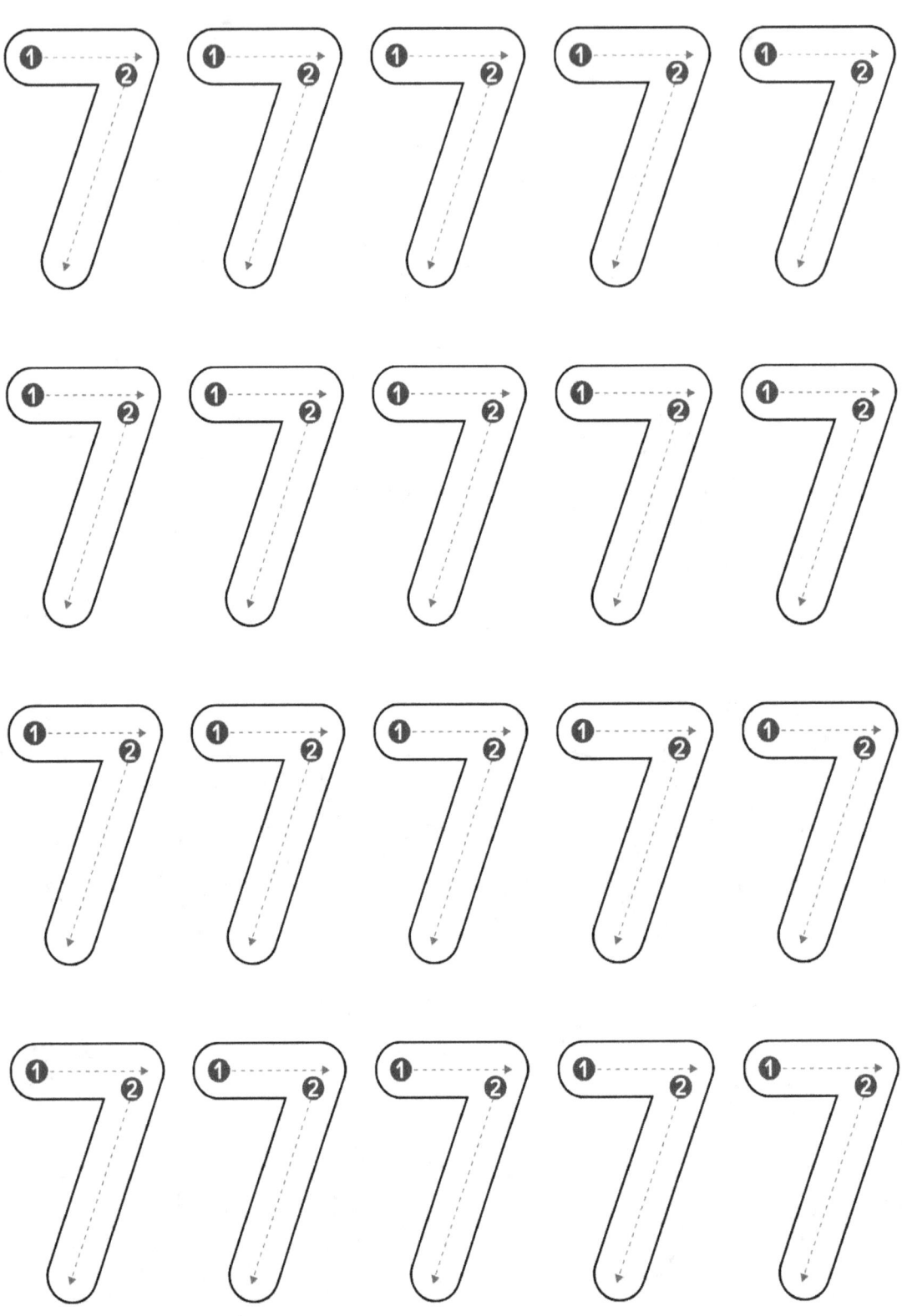

NUMBER TRACING

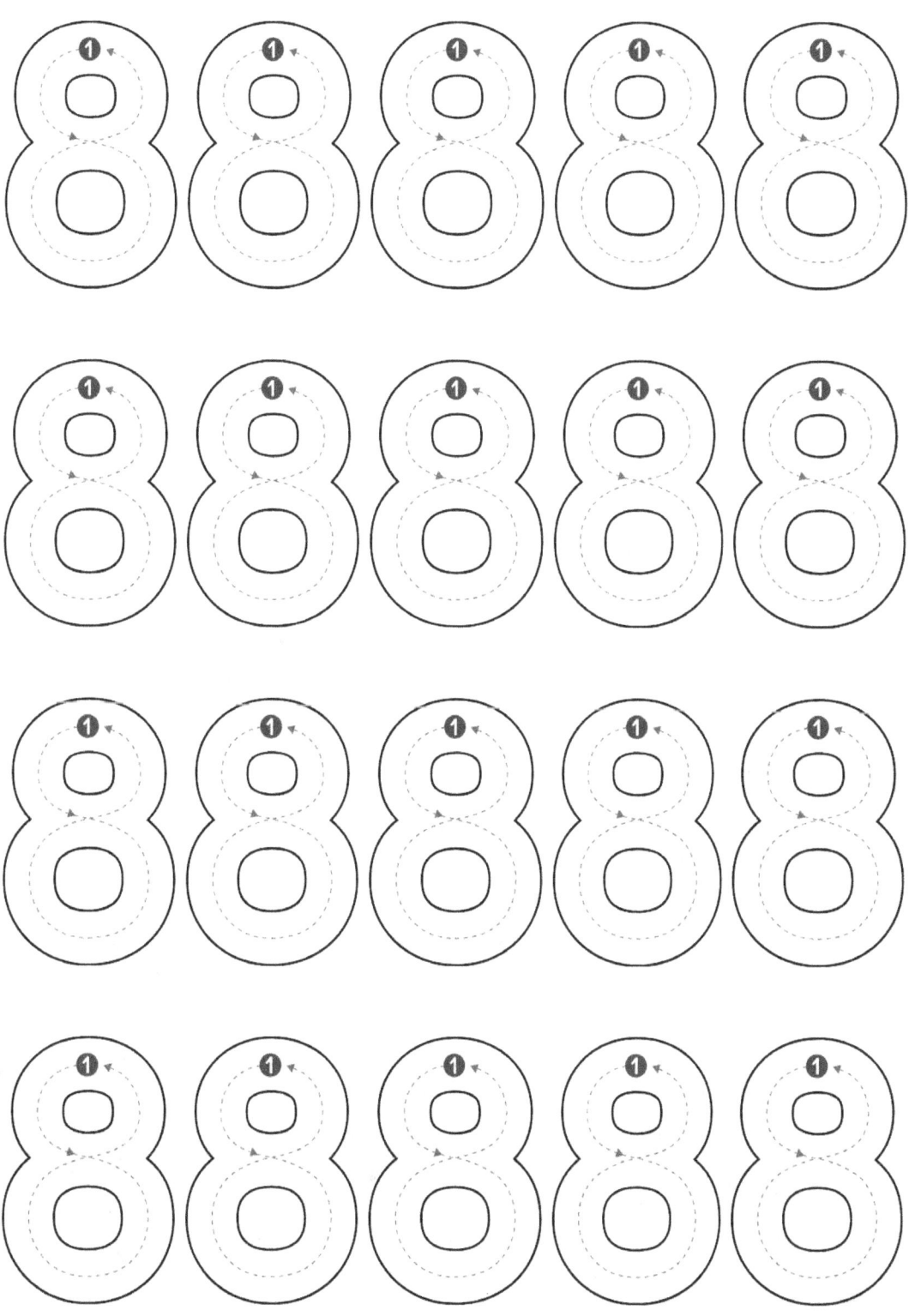

NUMBER TRACING

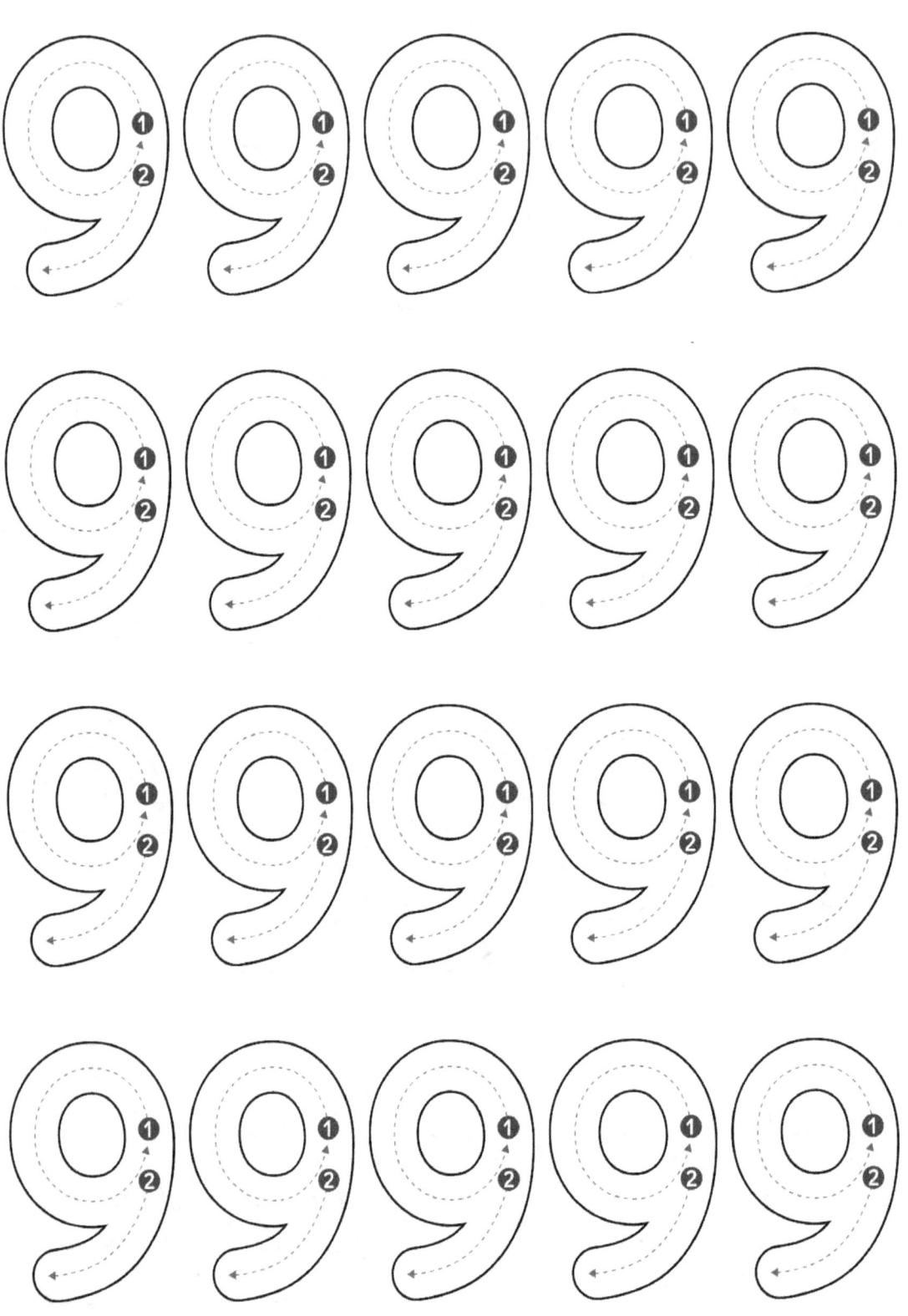

NUMBER TRACING

NUMBER TRACING

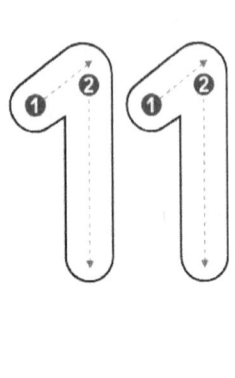 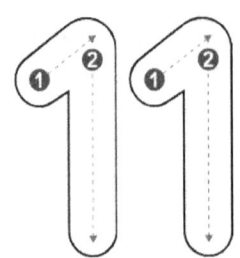 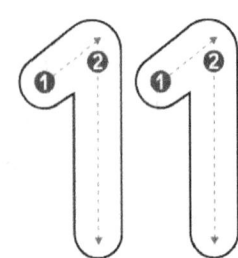

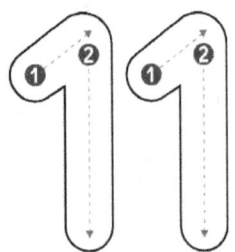 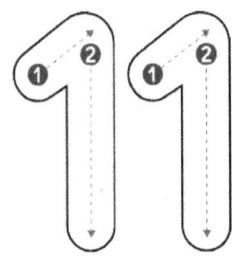 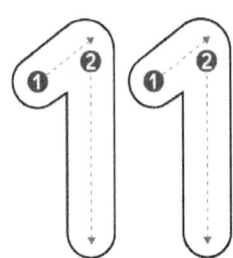

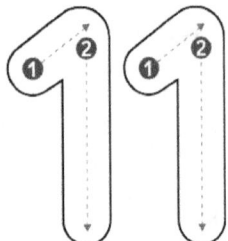 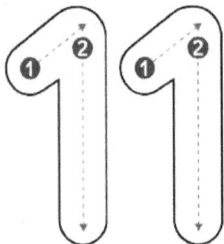 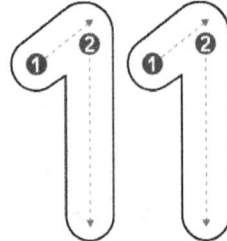

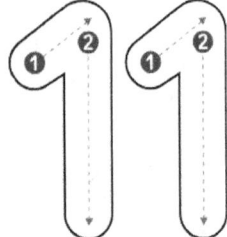 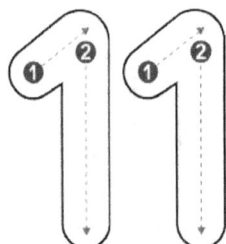

NUMBER TRACING

 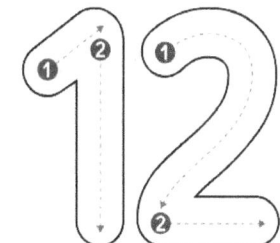 

NUMBER TRACING

NUMBER TRACING

NUMBER TRACING

NUMBER TRACING

NUMBER TRACING

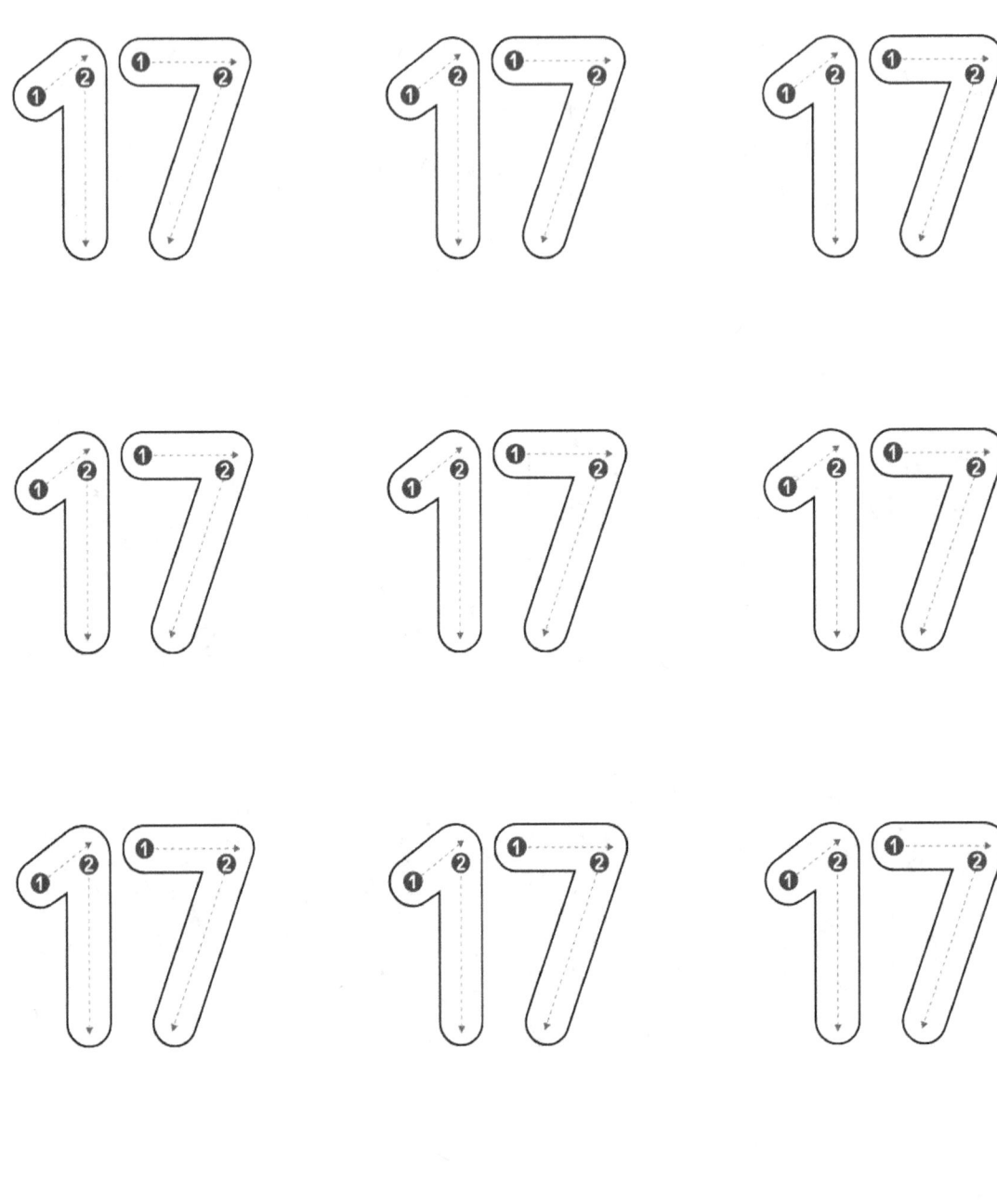

NUMBER TRACING

NUMBER TRACING

NUMBER TRACING

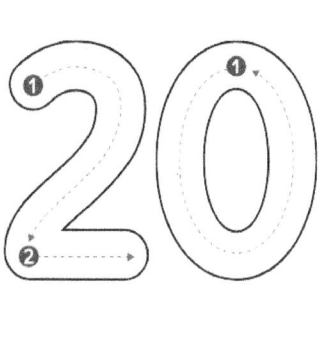  

 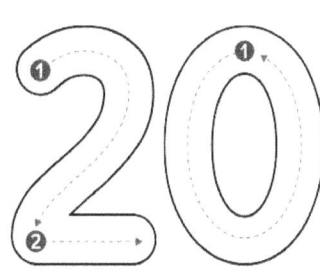 

 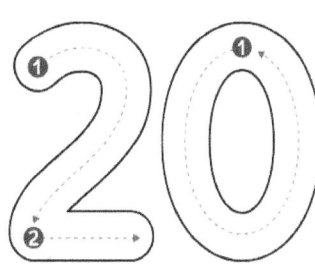 

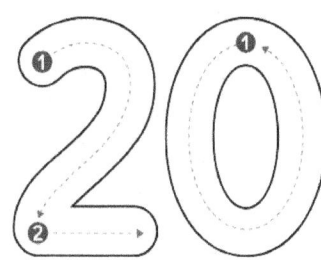 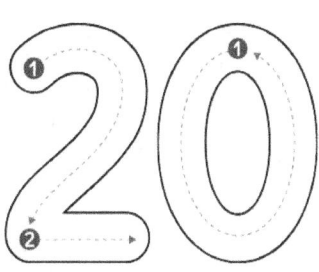 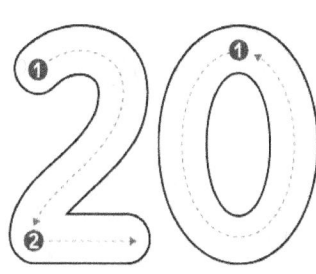

NUMBER TRACING

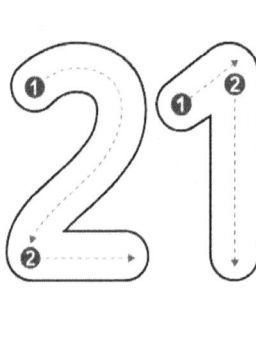  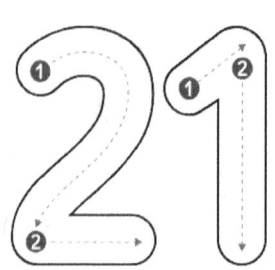

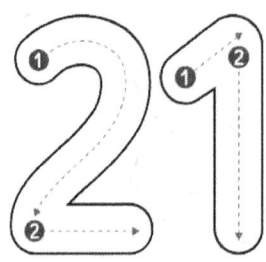  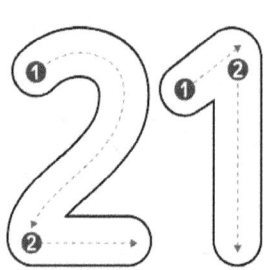

  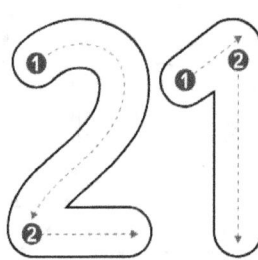

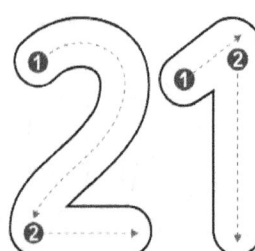 

NUMBER TRACING

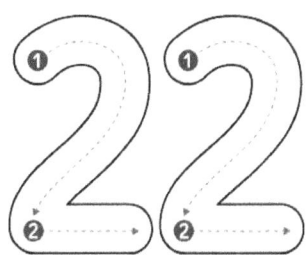 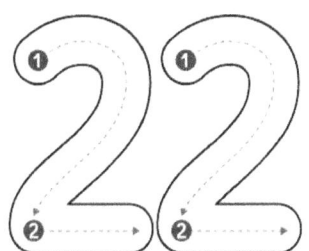 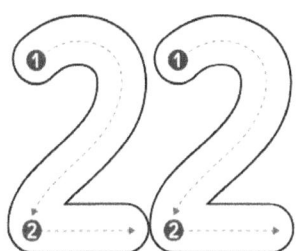

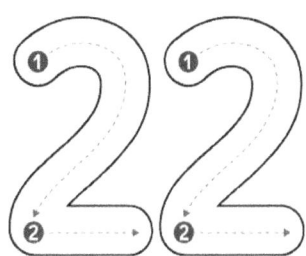 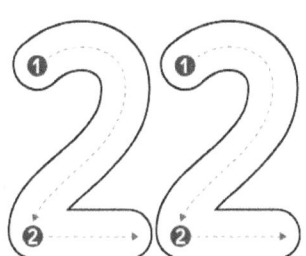 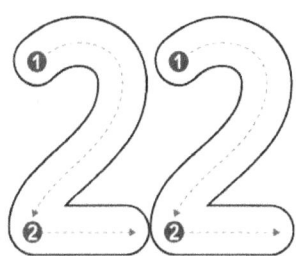

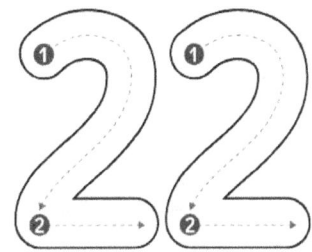 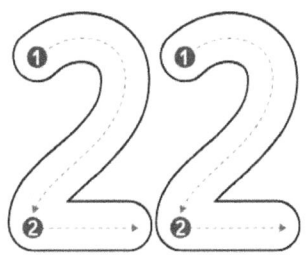 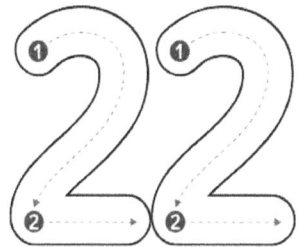

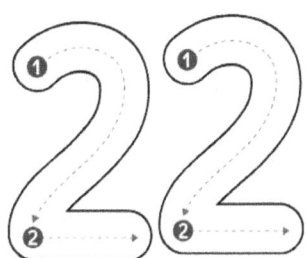 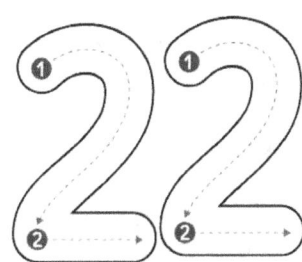 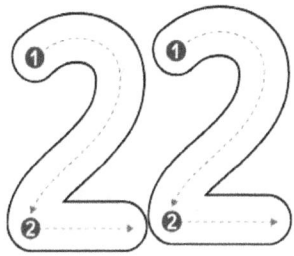

NUMBER TRACING

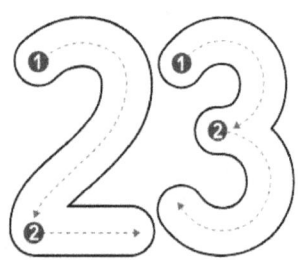  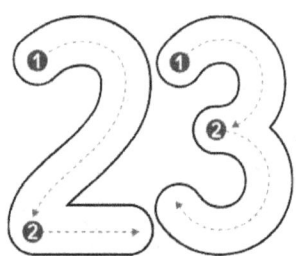

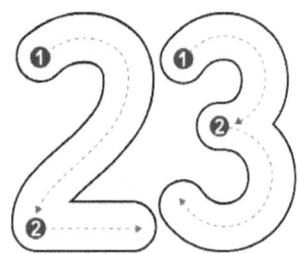 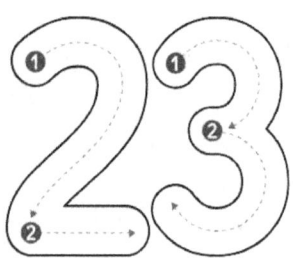 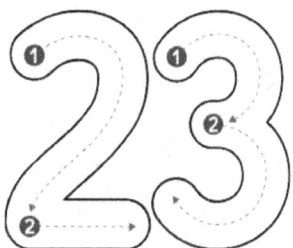

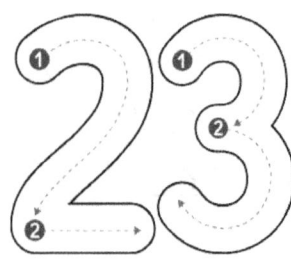  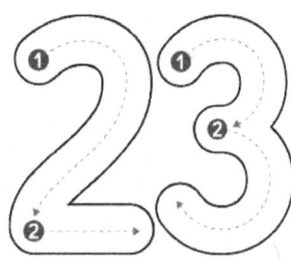

NUMBER TRACING

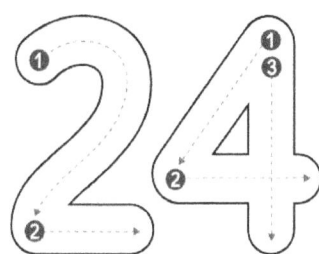 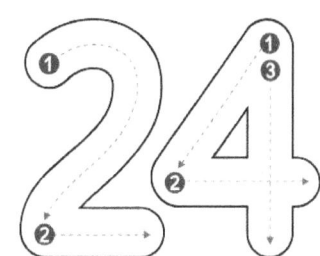 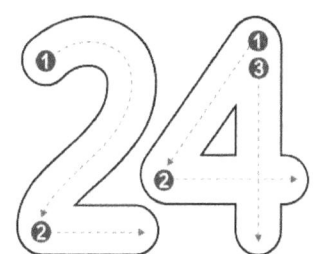

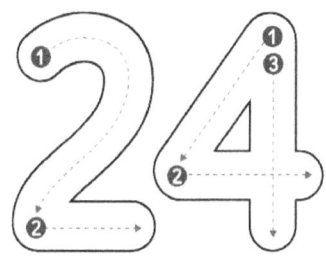 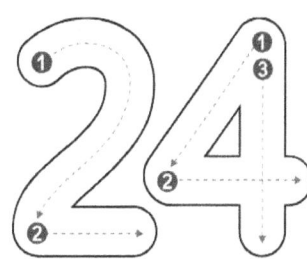 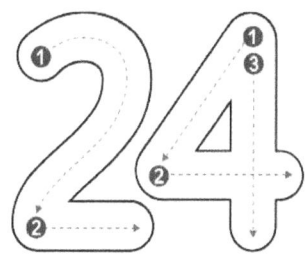

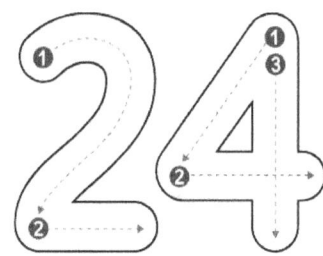 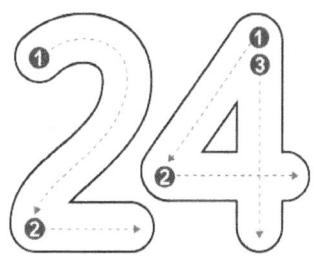 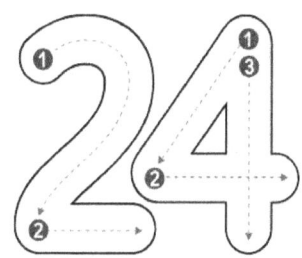

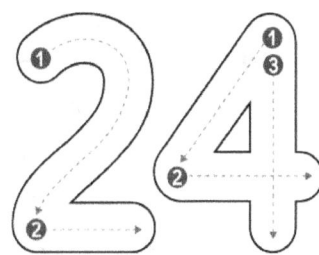 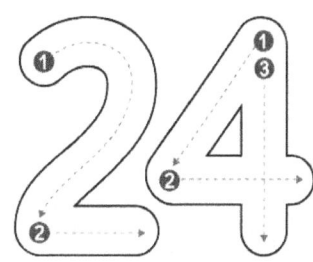 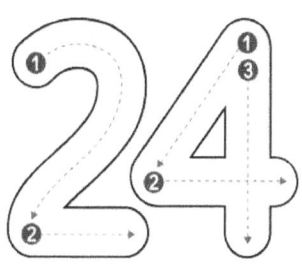

NUMBER TRACING

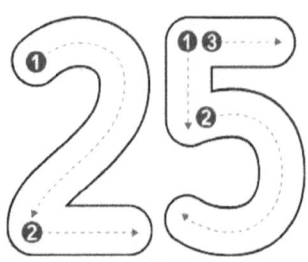 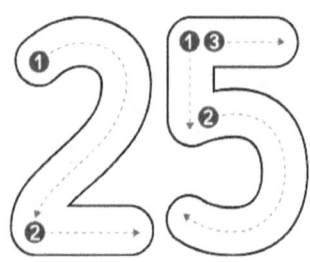 

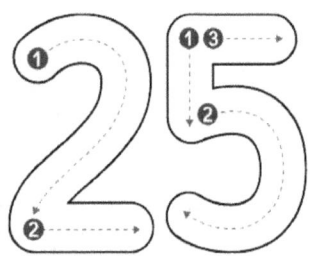 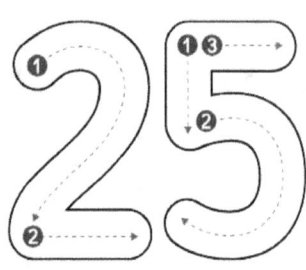 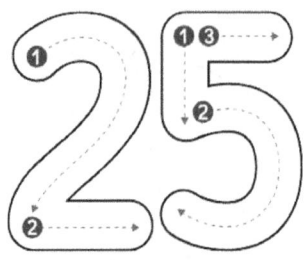

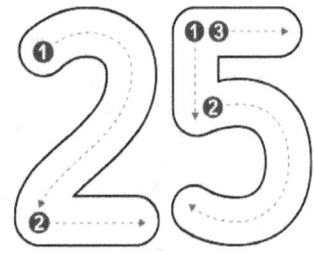 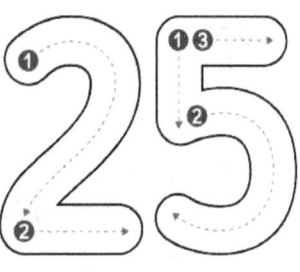 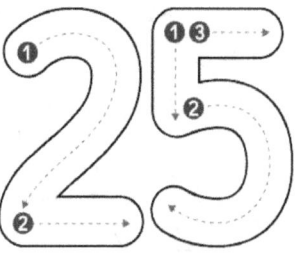

 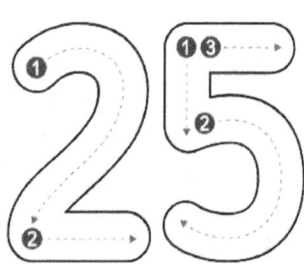 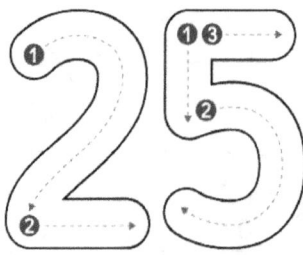

NUMBER TRACING

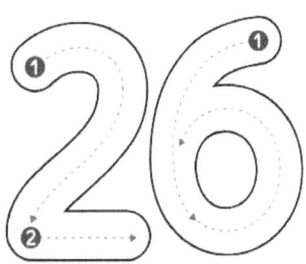  

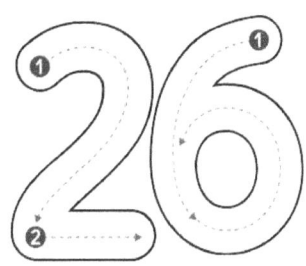 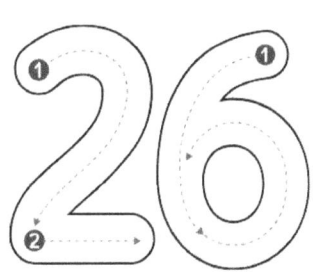 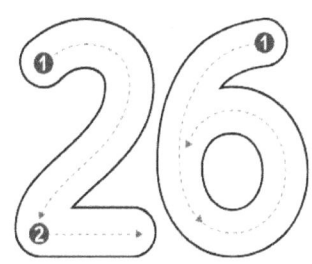

  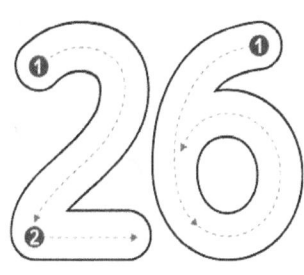

NUMBER TRACING

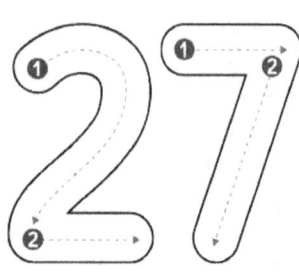  

 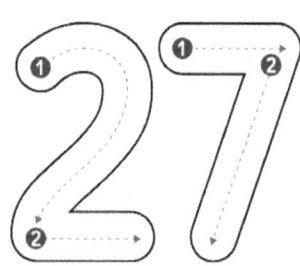 

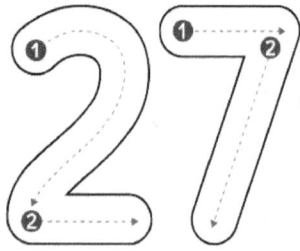 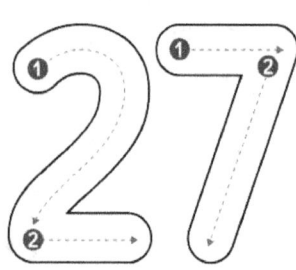 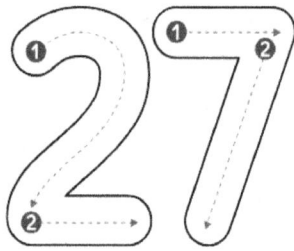

 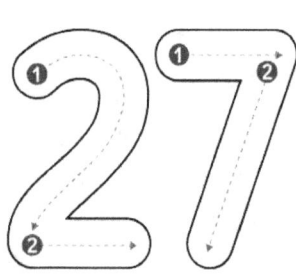 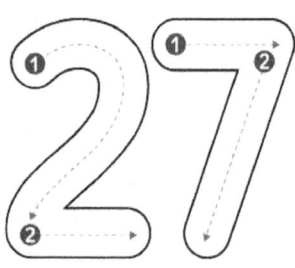

NUMBER TRACING

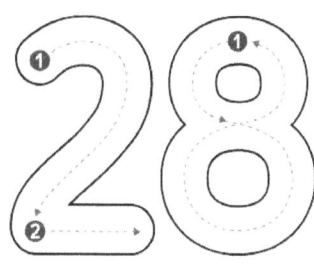 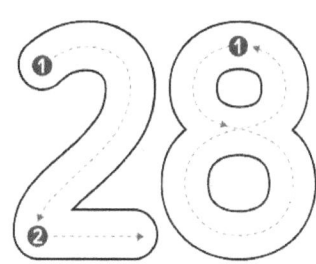 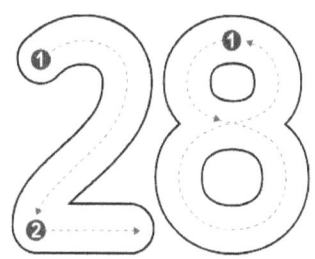

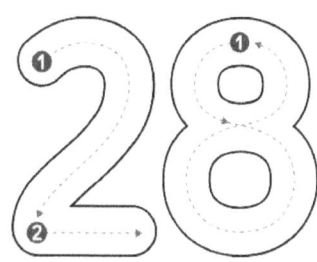 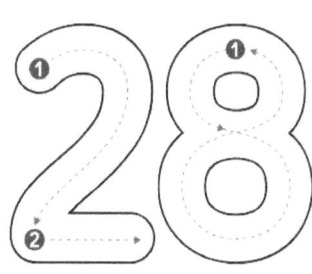 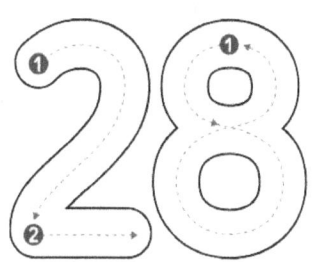

 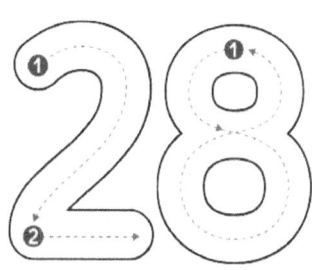 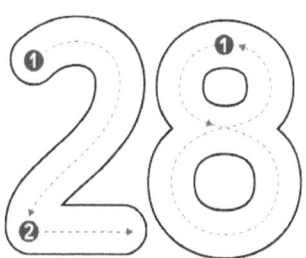

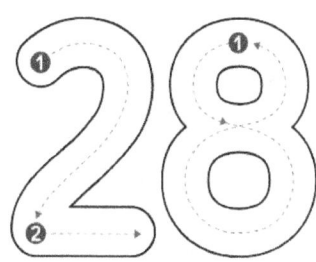 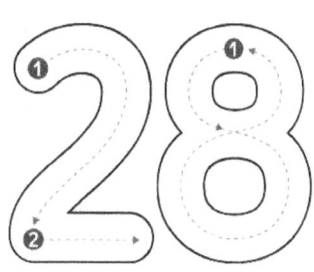 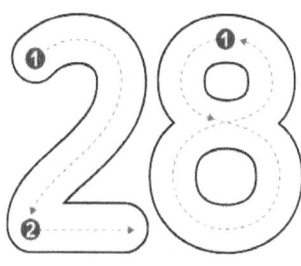

NUMBER TRACING

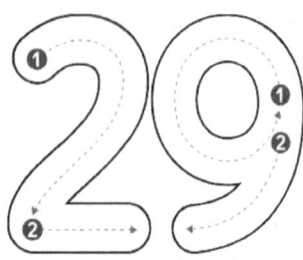 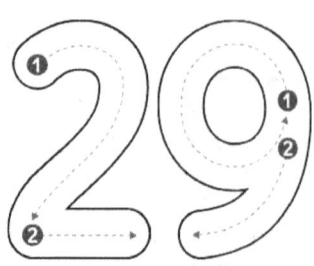 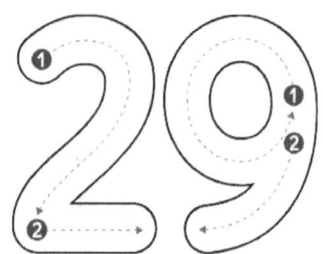

 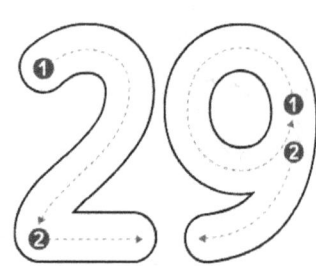 

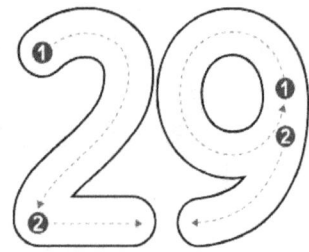 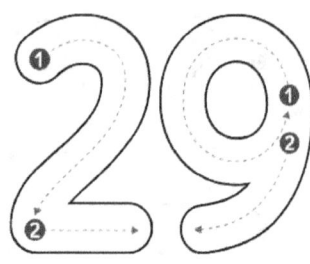 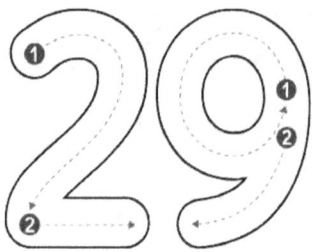

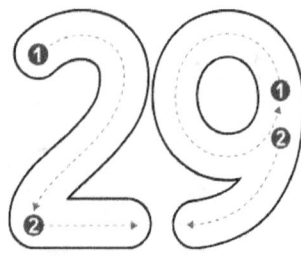 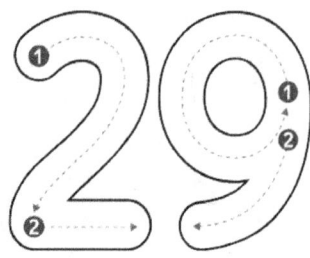 

NUMBER TRACING

LET'S COUNT

Count the different objects and write your answers in the chart below

NUMBER AFTER

Fill in the number that comes after:

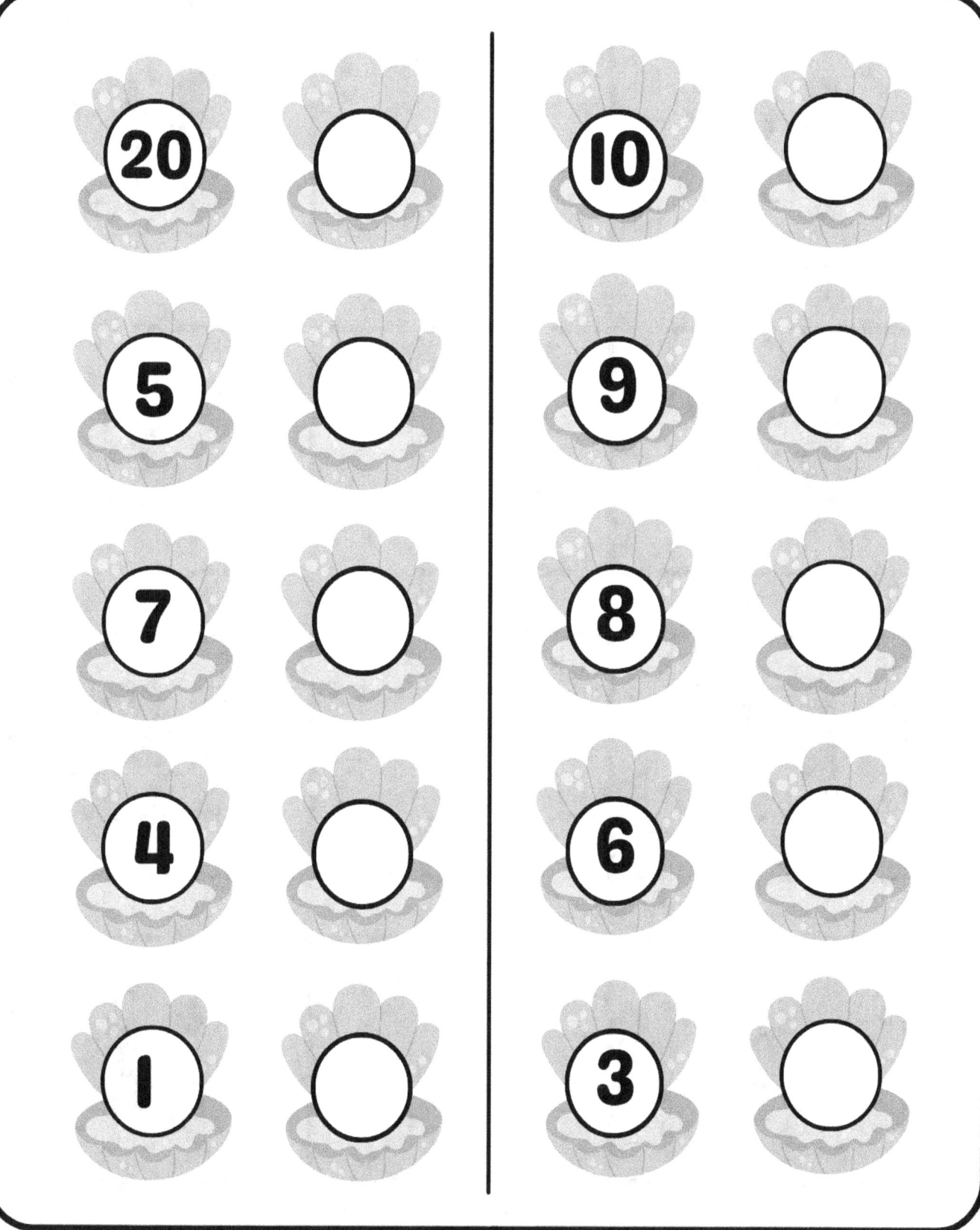

NUMBER BETWEEN

Fill in the number that comes between to complete the sequence.

18		20

16		18

13		15

1		3

COUNT BY 2'S

Fill in the numbers to complete the sequence.

13   15   ☐   19   21

23   25   27   29   ☐

☐   35   37   39   41

43   45   ☐   ☐   51

COLOR THE DOTS

Count out and color the number of
dots for each number below.

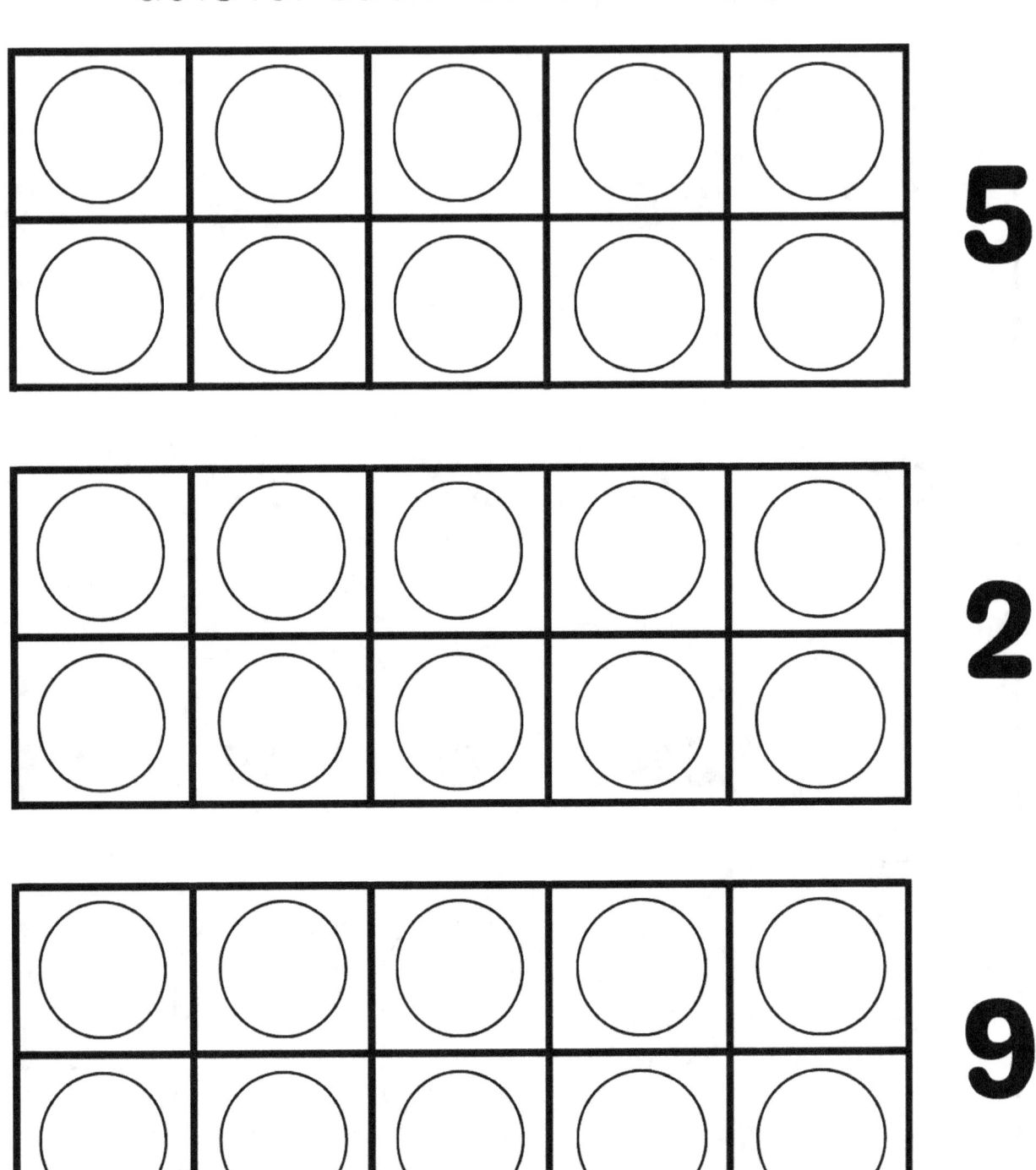

COUNT AND COLOR

Count the images and color the correct answer

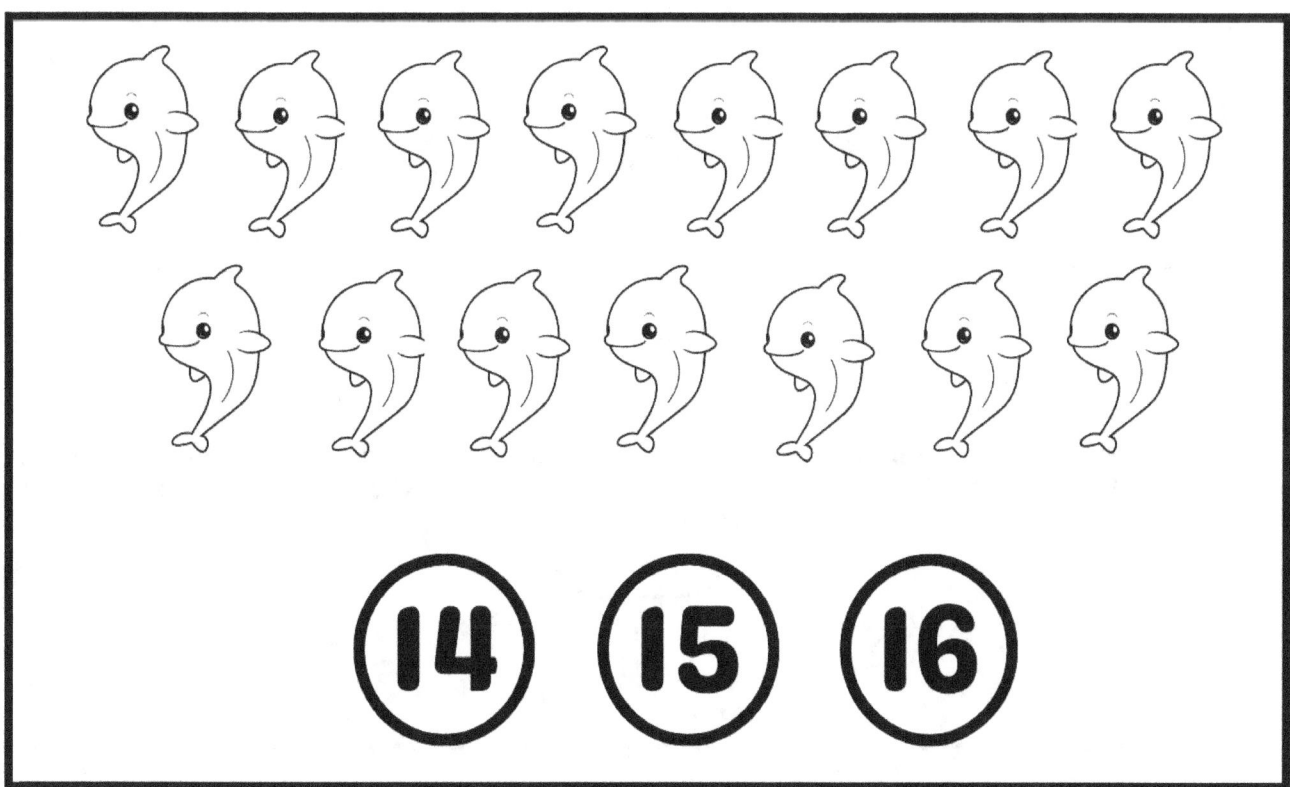

COUNT FORWARD

Fill in the numbers to complete the sequence.

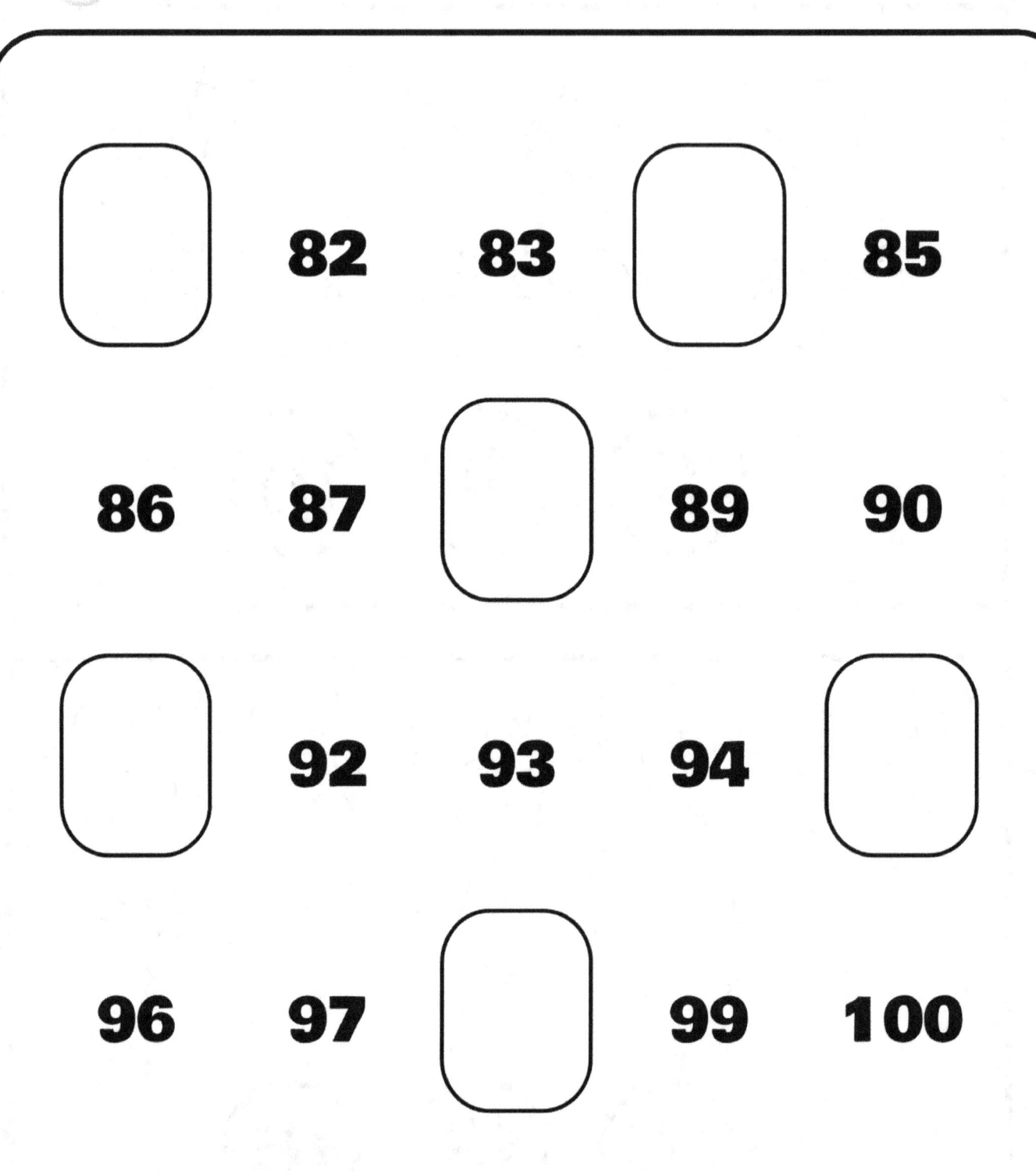

82 83 ____ 85

86 87 ____ 89 90

____ 92 93 94 ____

96 97 ____ 99 100

NUMBER RECOGNITION

Find the numbers and color in with the color given.

1 Light Blue **2** Pink **3** Green **4** Dark Blue

LET'S COUNT

Count the different objects and write your answers in the chart below

NUMBER AFTER

Fill in the number that comes after:

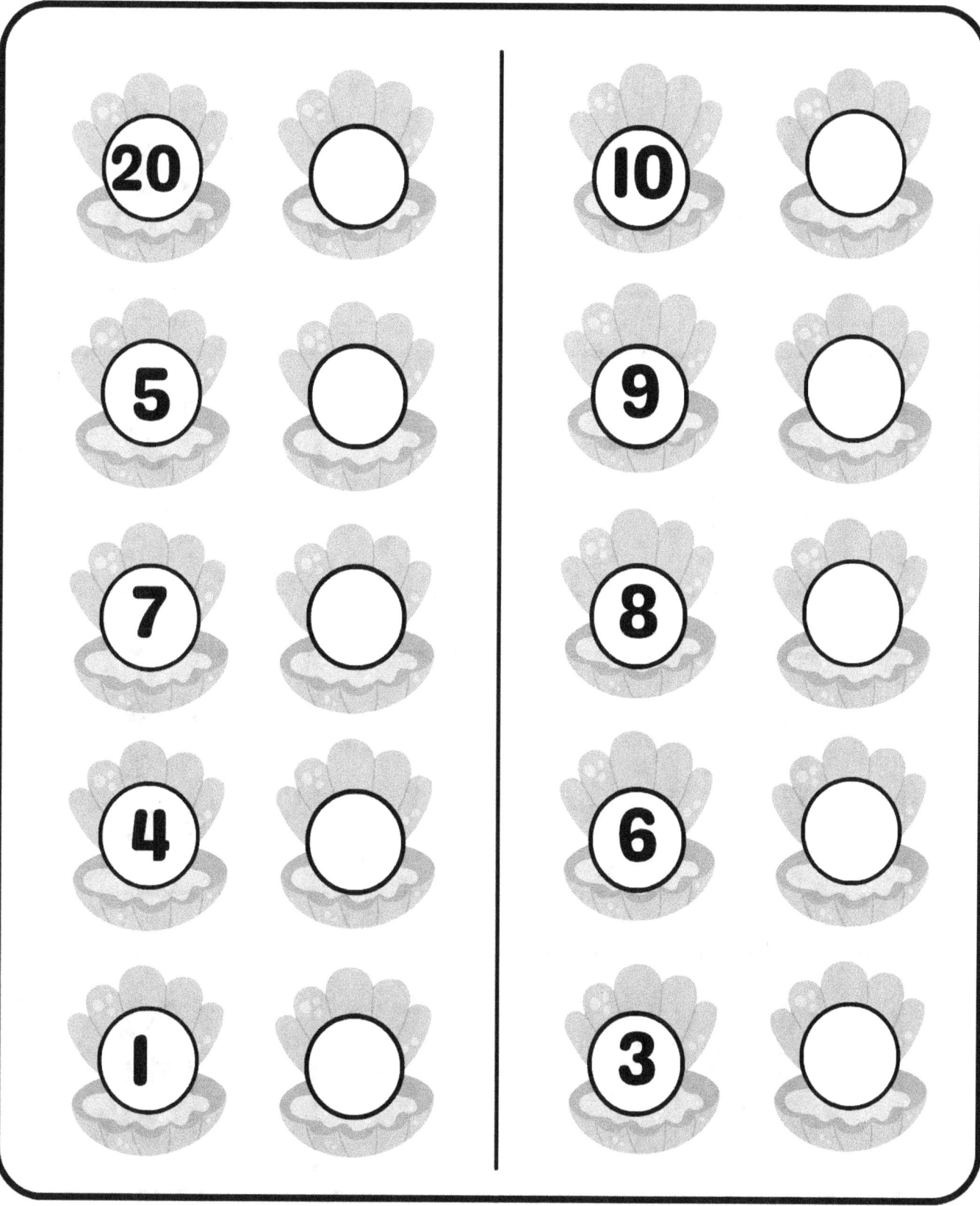

NUMBER BETWEEN

Fill in the number that comes between to
complete the sequence.

9		11

5		7

0		2

4		6

COUNT BY 2'S

Fill in the numbers to complete the sequence.

17   19   21   [   ]   25

27   [   ]   [   ]   33   [   ]

37   39   41   43   45

47   49   51   [   ]   55

COUNT AND COLOR

Count the images and color the correct answer

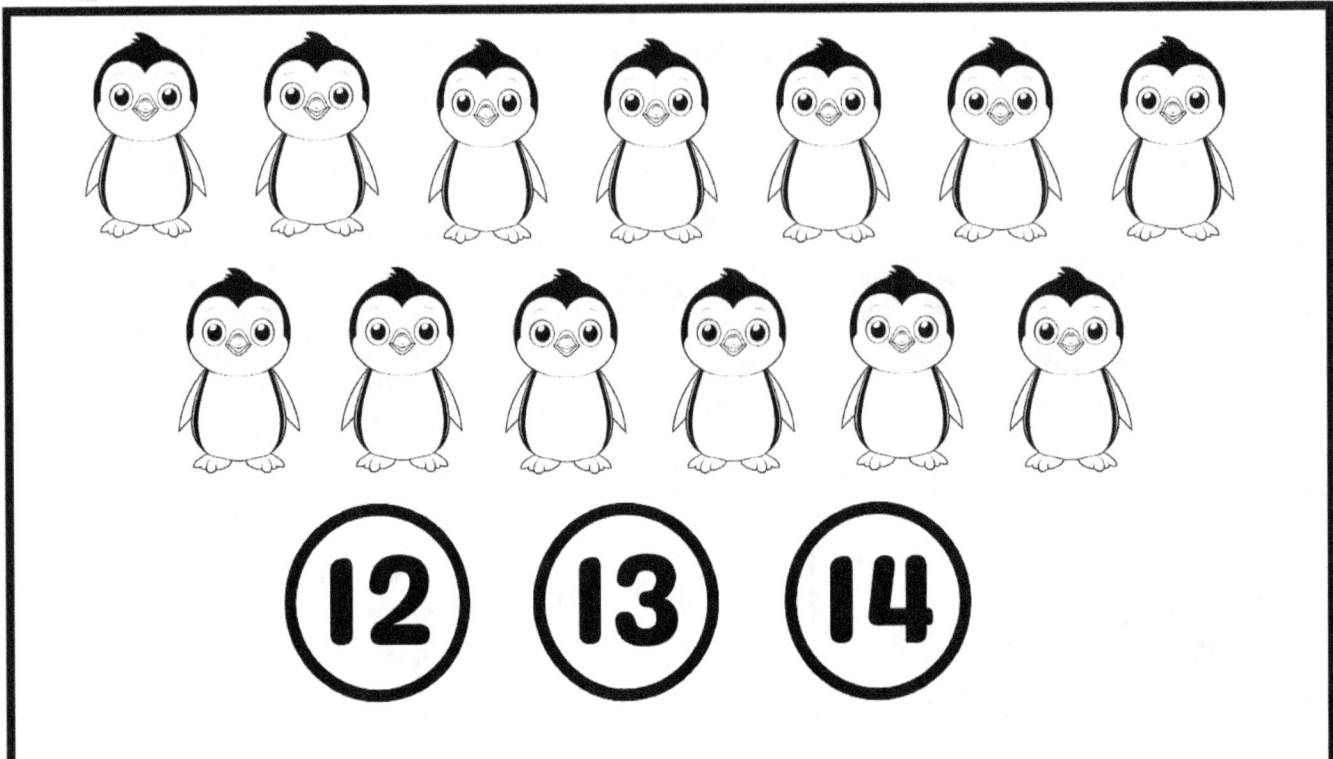

COUNT FORWARD

Fill in the numbers to complete the sequence.

	18	19		21
22	23		25	26
	28	29	30	
32	33		35	36

COLOR THE DOTS

Count out and color the number of
dots for each number below.

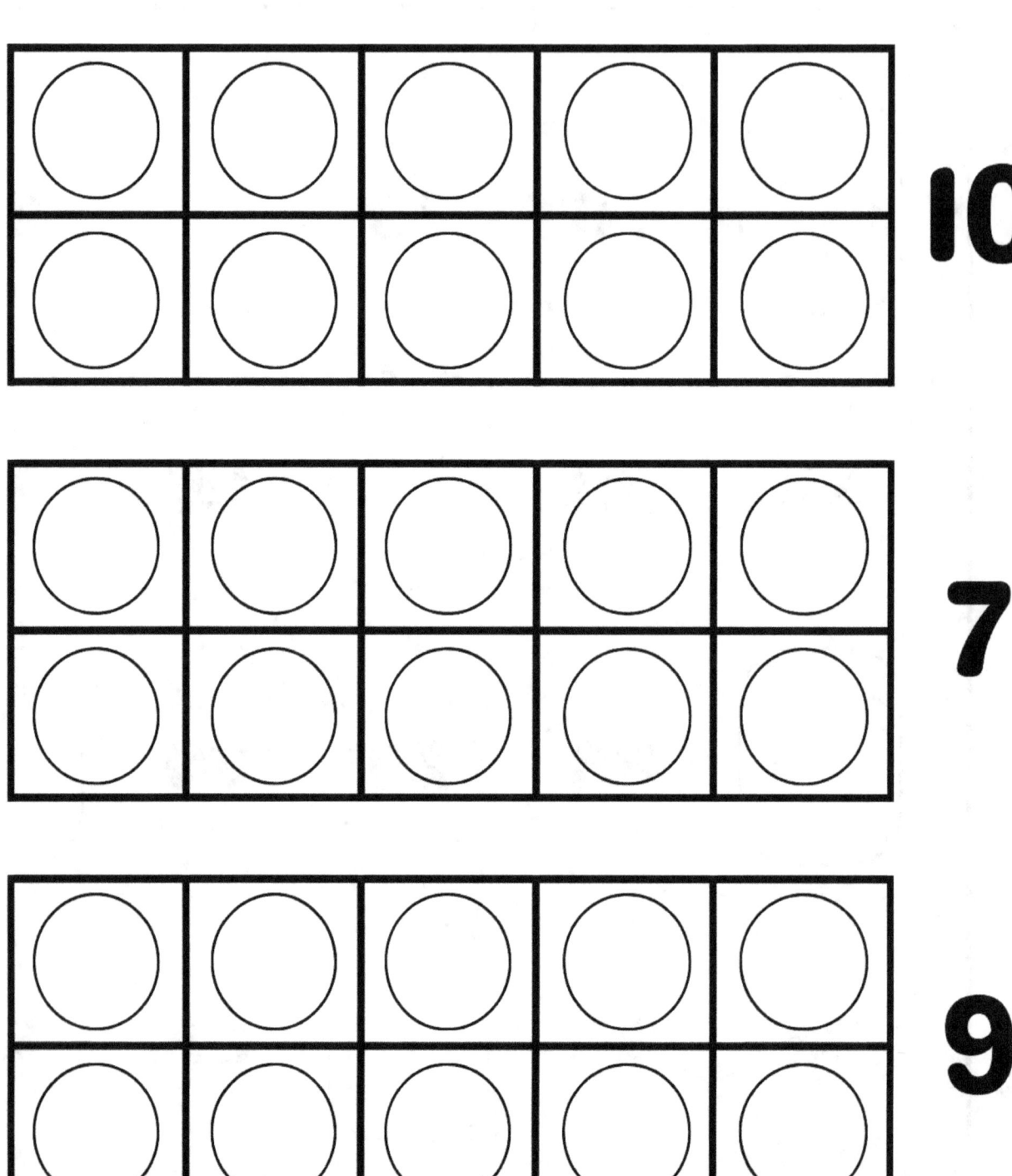

COUNT AND COLOR

19	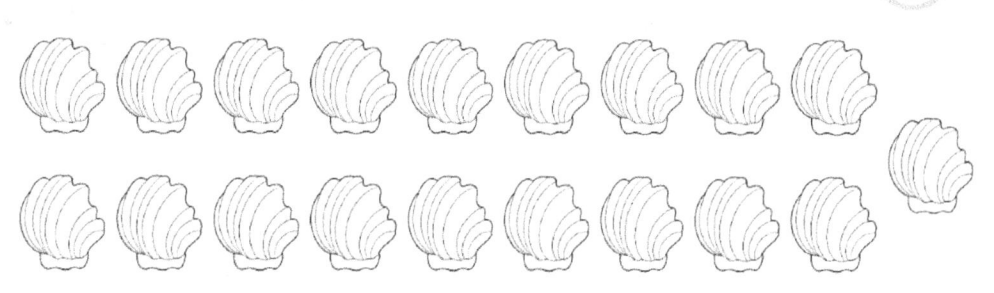
20	
18	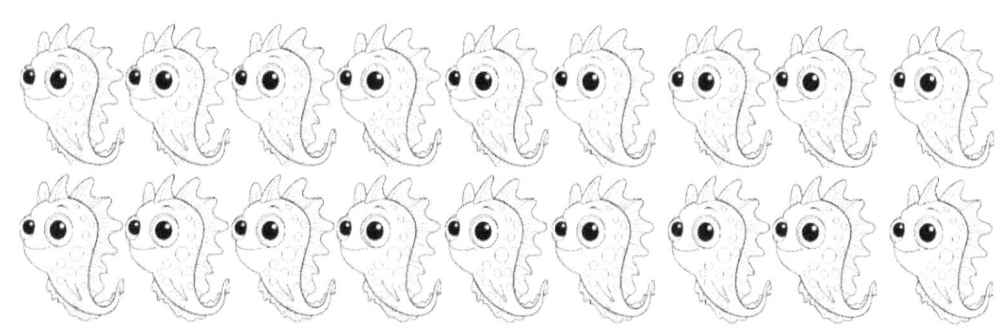
17	

LET'S COUNT

Count the different objects and write your answers in the chart below

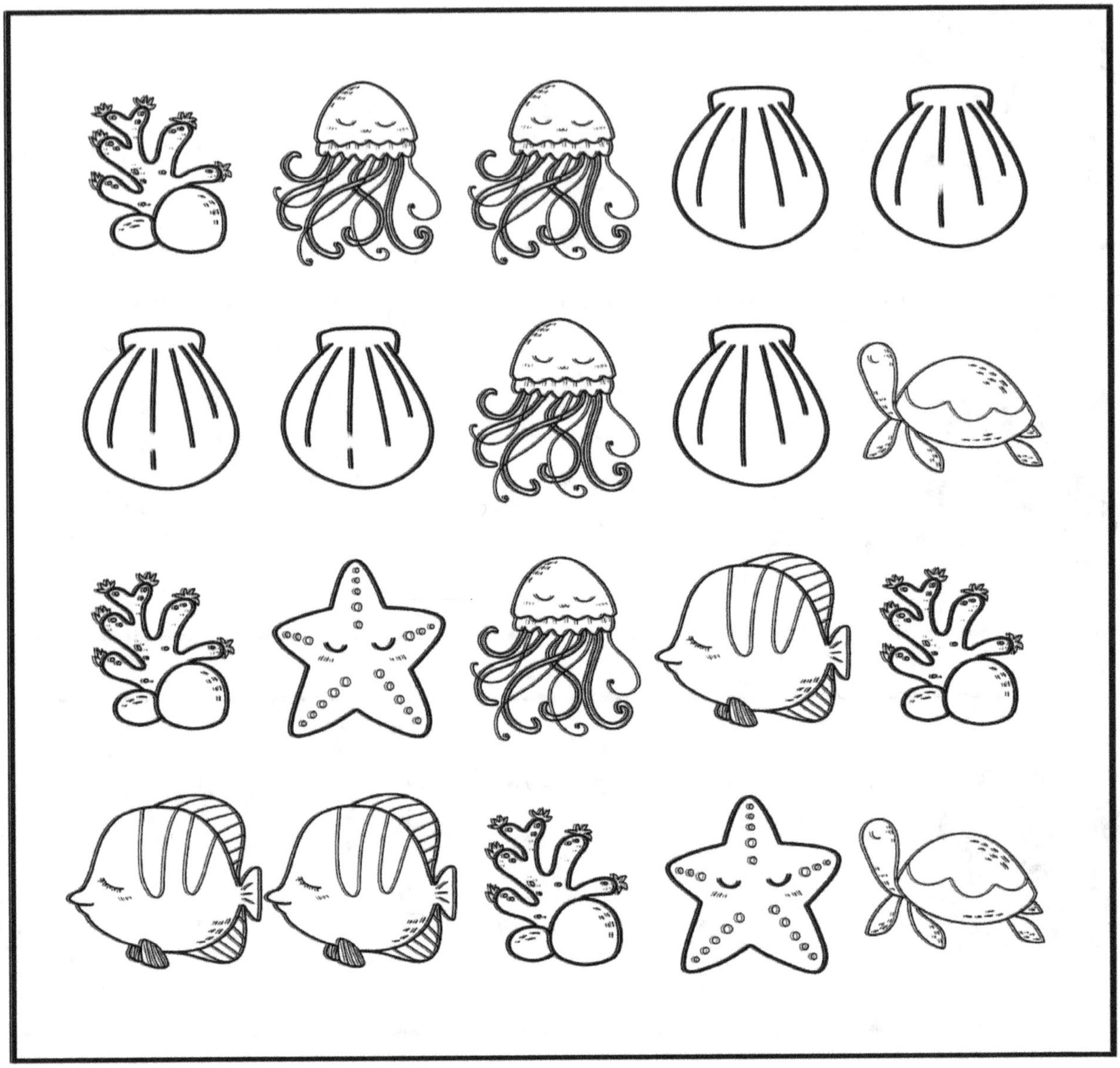

NUMBER BEFORE

Fill in the number that comes before:

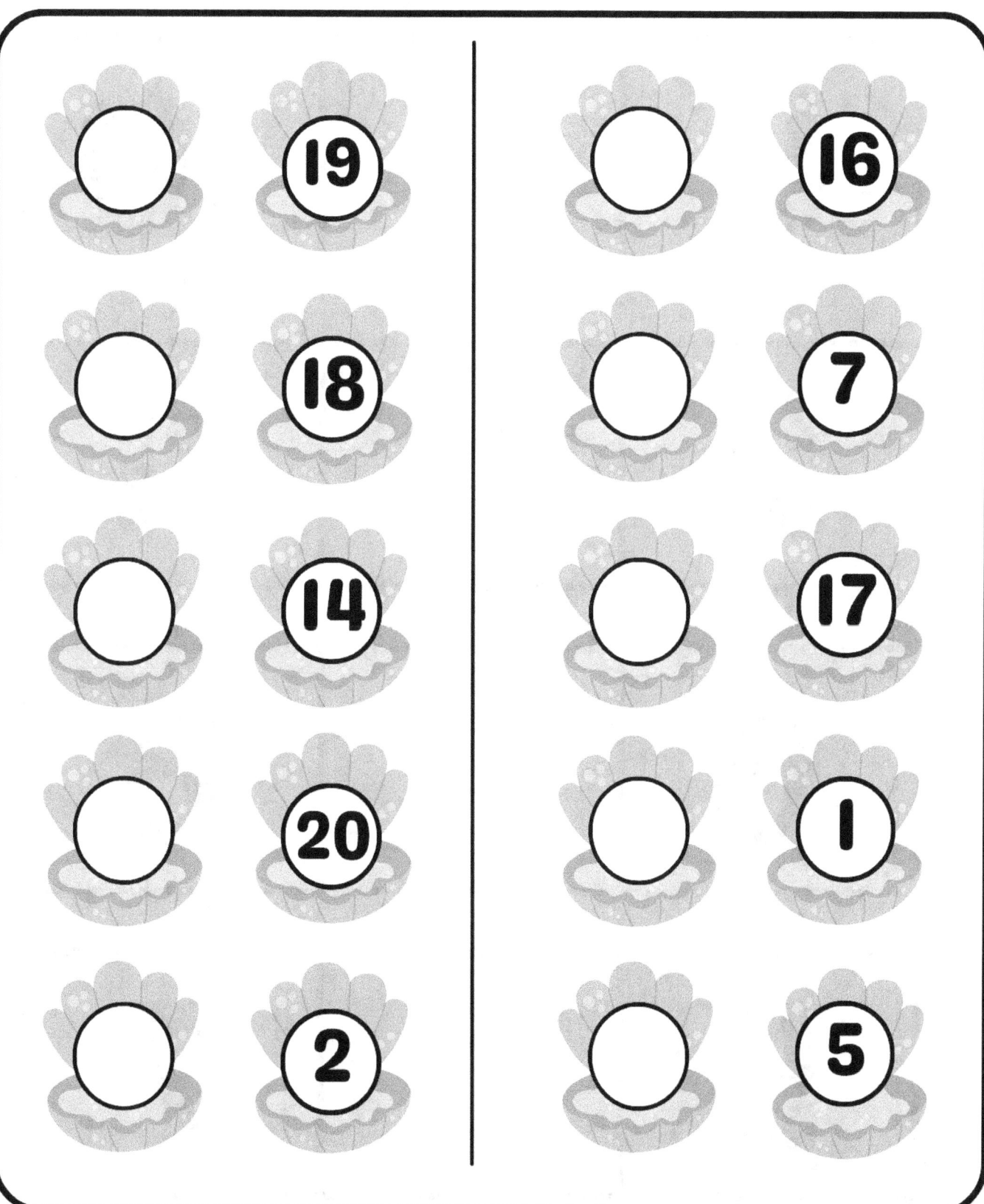

___ 19

___ 18

___ 14

___ 20

___ 2

___ 16

___ 7

___ 17

___ 1

___ 5

NUMBER BETWEEN

Fill in the number that comes between to complete the sequence.

8		10

2		4

6		8

5		7

COUNT BY 2'S

Fill in the numbers to complete the sequence.

17   19   21   [ ]   25

27   [ ]   [ ]   33   [ ]

37   39   41   43   45

47   49   51   [ ]   55

COLOR THE DOTS

Count out and color the number of dots for each number below.

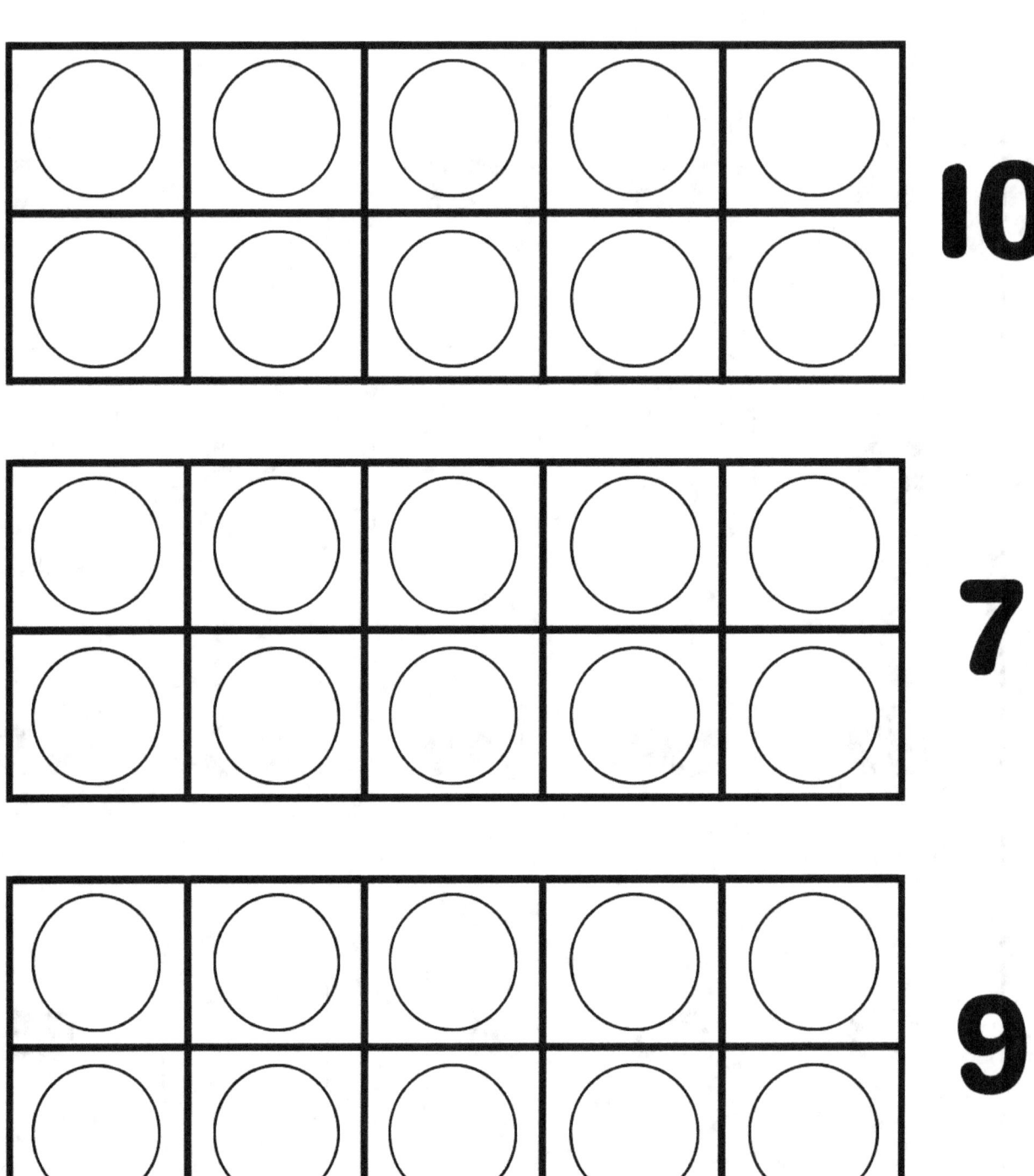

10

7

9

COUNT AND COLOR

Count the images and color the correct answer

COUNT FORWARD

Fill in the numbers to complete the sequence.

	18	19		21
22	23		25	26
	28	29	30	
32	33		35	36

NUMBER RECOGNITION

Find the numbers and color in with the color given.

1 Blue **2** Pink **3** Yellow **4** Green **5** Brown

COUNT AND COLOR

11	

9	

12	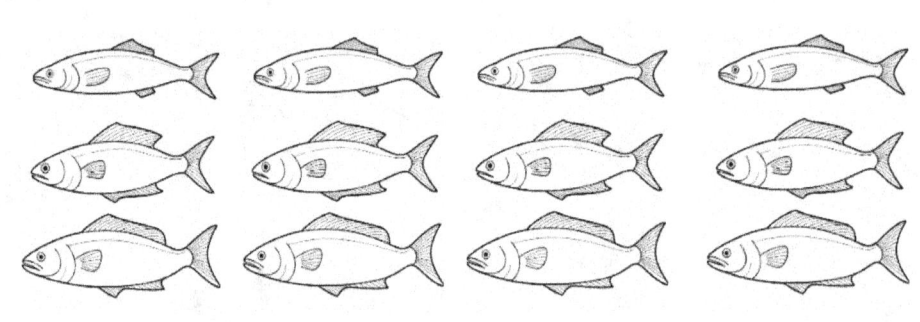

10	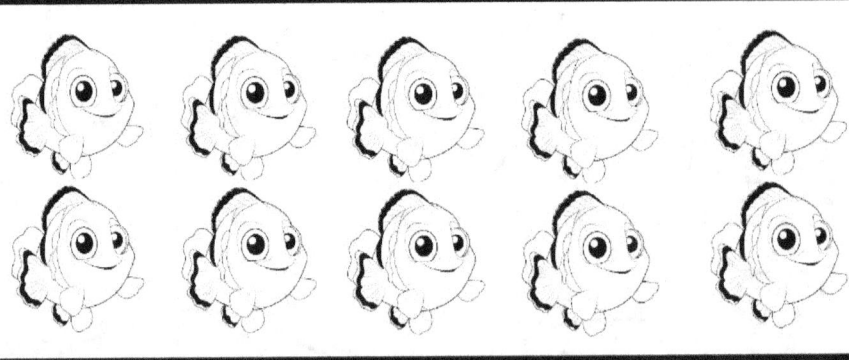

MISSING NUMBERS

Help Finley find his cave by filling in the missing numbers.

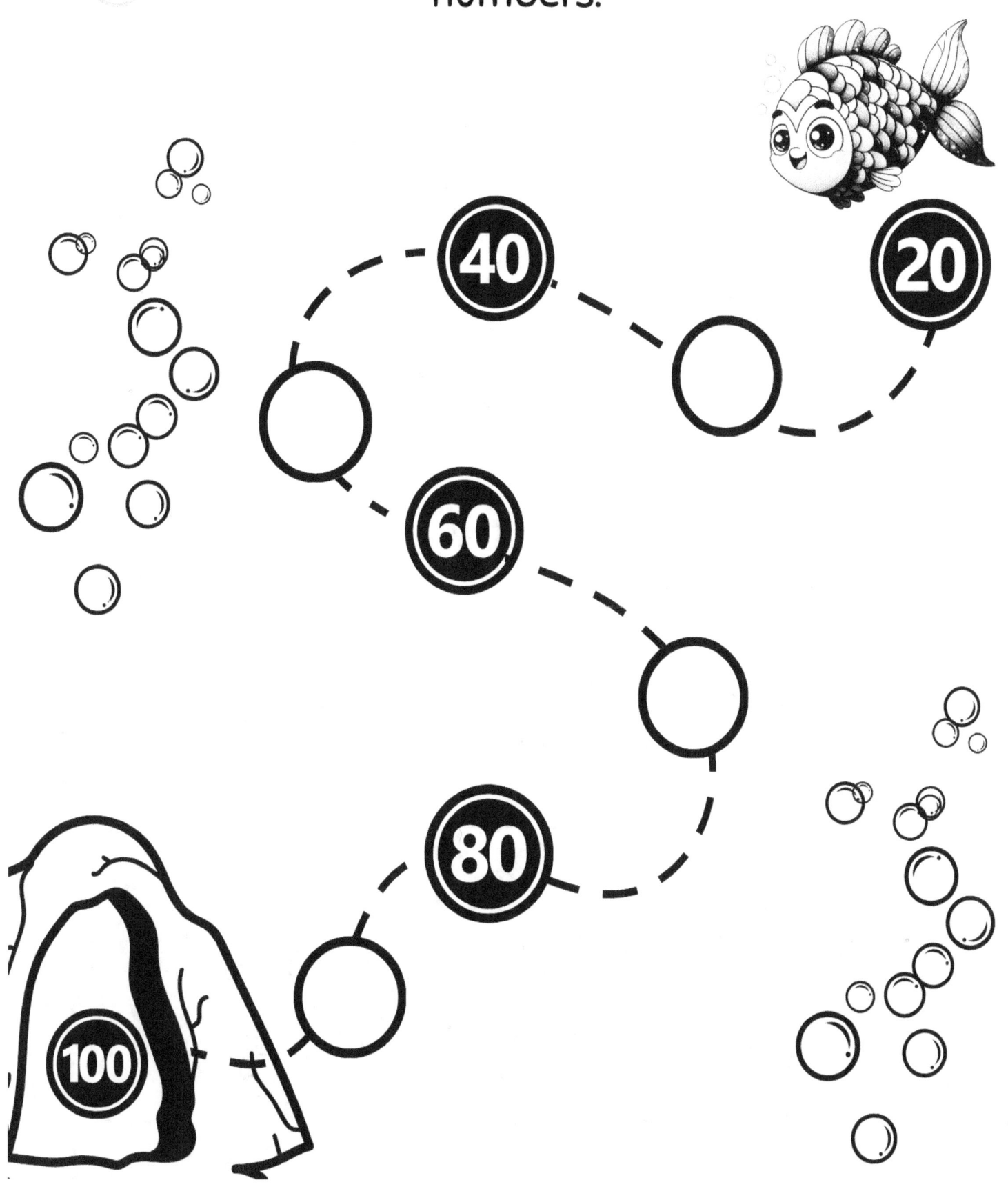

LET'S COUNT

Count the different objects and write your
answers in the chart below

NUMBER BEFORE

Fill in the number that comes before:

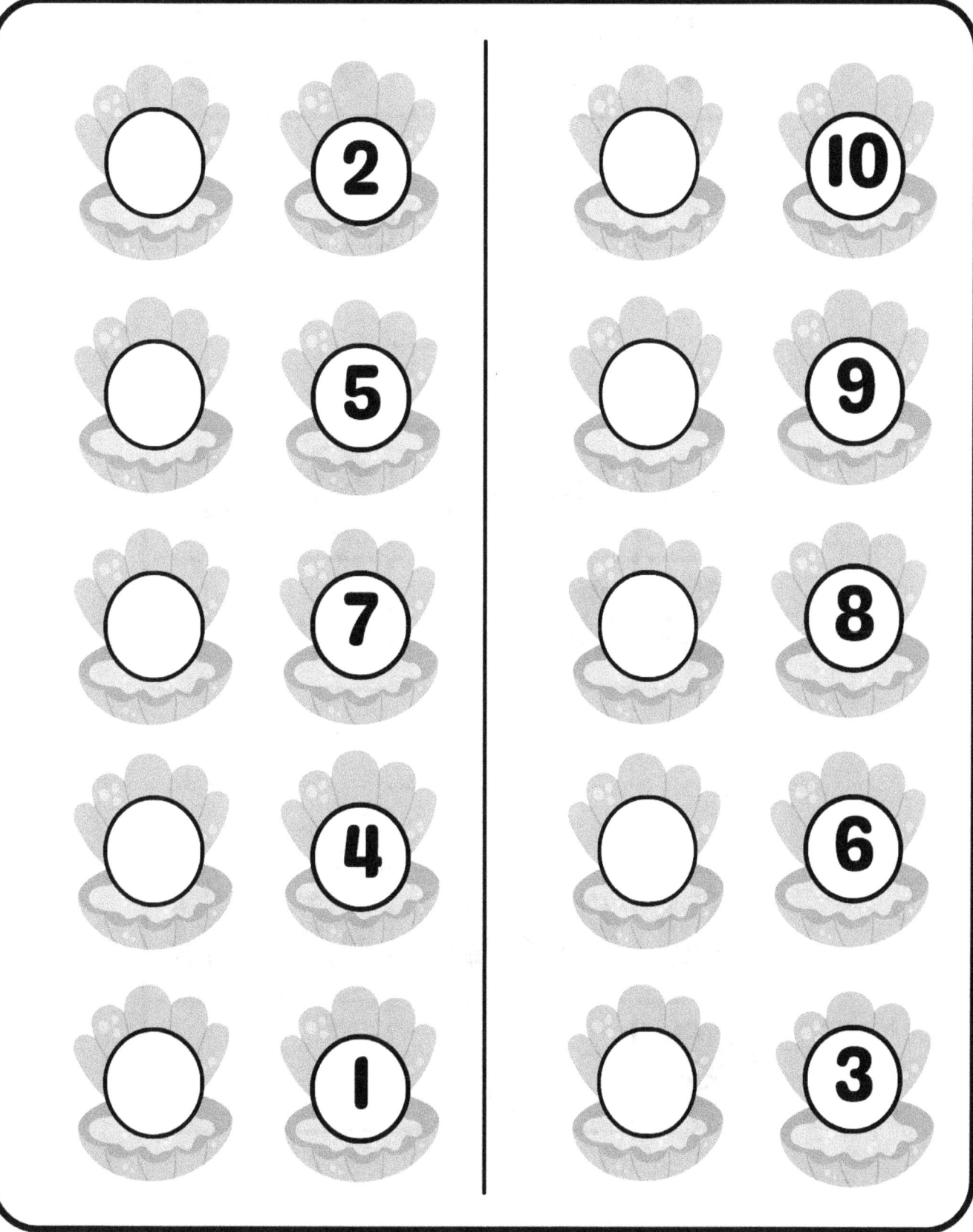

COUNT THE DOTS

Count the dots and write your answer.

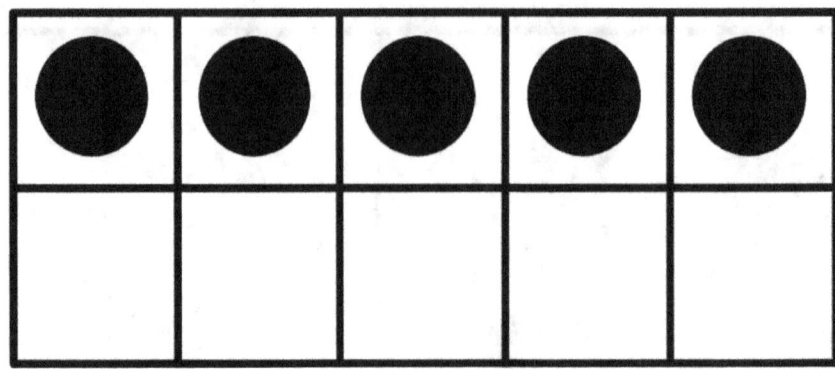

———

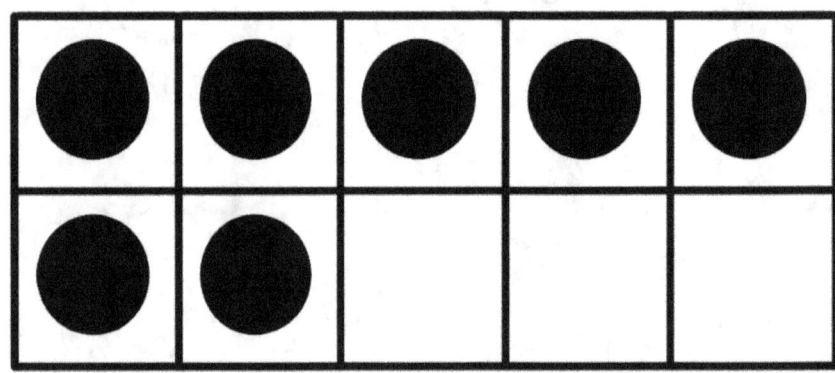

———

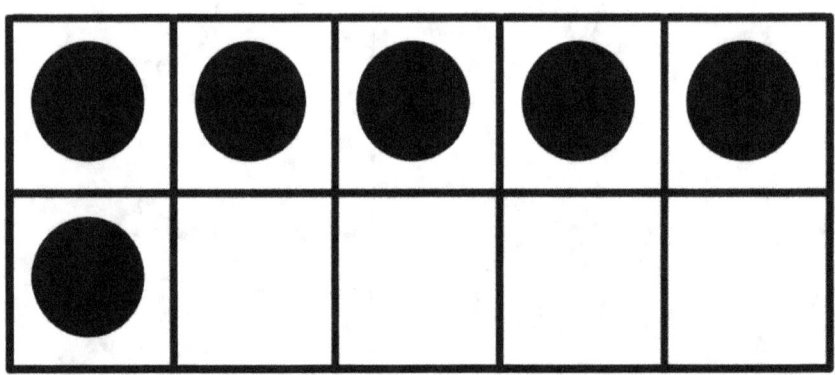

———

NUMBER BETWEEN

Fill in the number that comes between to complete the sequence.

11		13

15		17

12		14

14		16

COLOR THE DOTS

Count out and color the number of
dots for each number below.

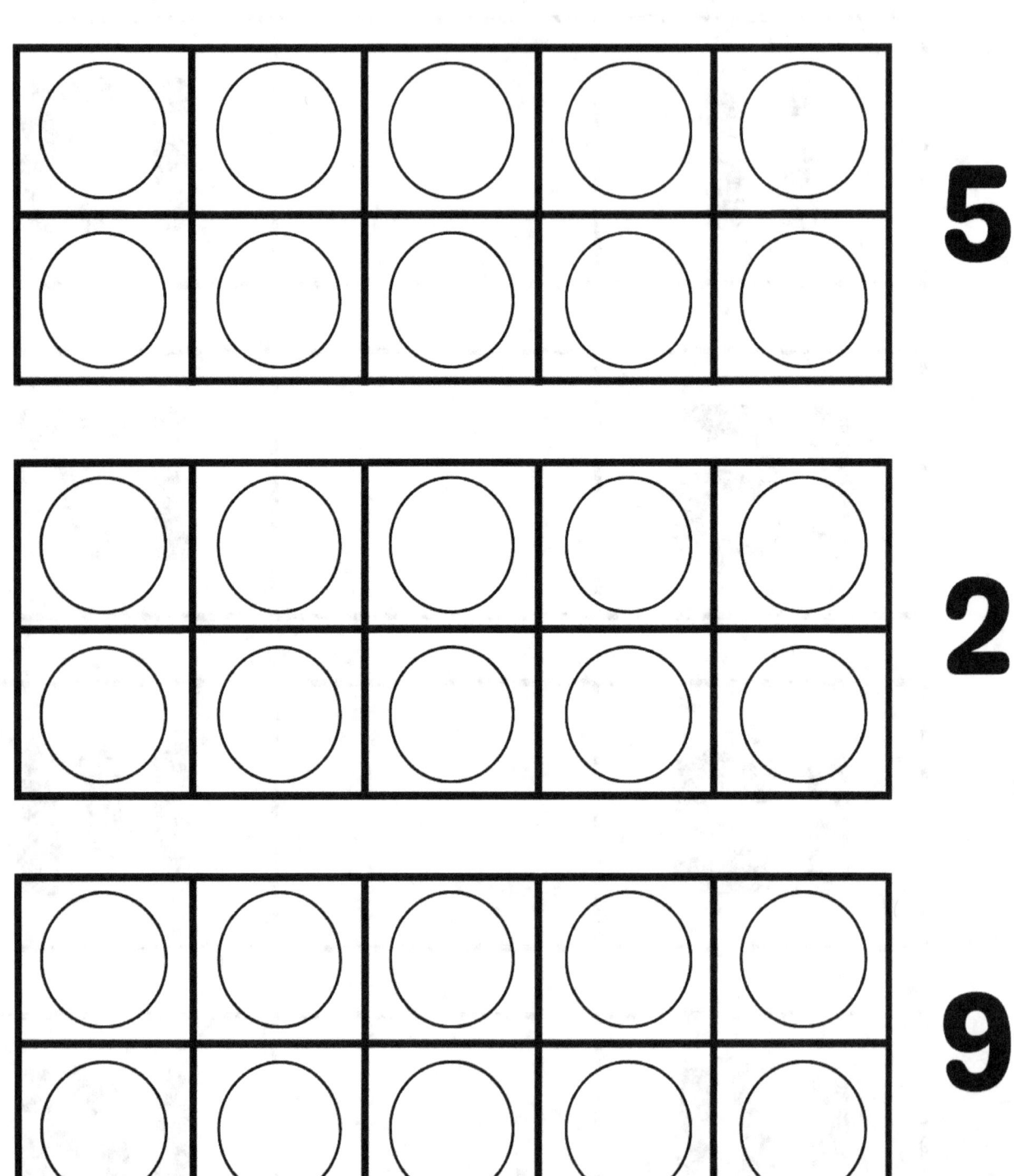

COUNT FORWARD

Fill in the numbers to complete the sequence.

[ ] 5   6 [ ]   8

9   10 [ ]   12   13

[ ]   15   16   17 [ ]

19   20 [ ]   22   23

NUMBER RECOGNITION

Find the numbers and color in with the color given.

6 Pink **7** Light Blue **8** Yellow **9** Green **10** Blue

COUNT AND COLOR

| 8 |  |

| 1 |  |

| 7 |  |

| 2 |   |

MISSING NUMBERS

Fill in the missing numbers to help Finley find Sammy Snail.

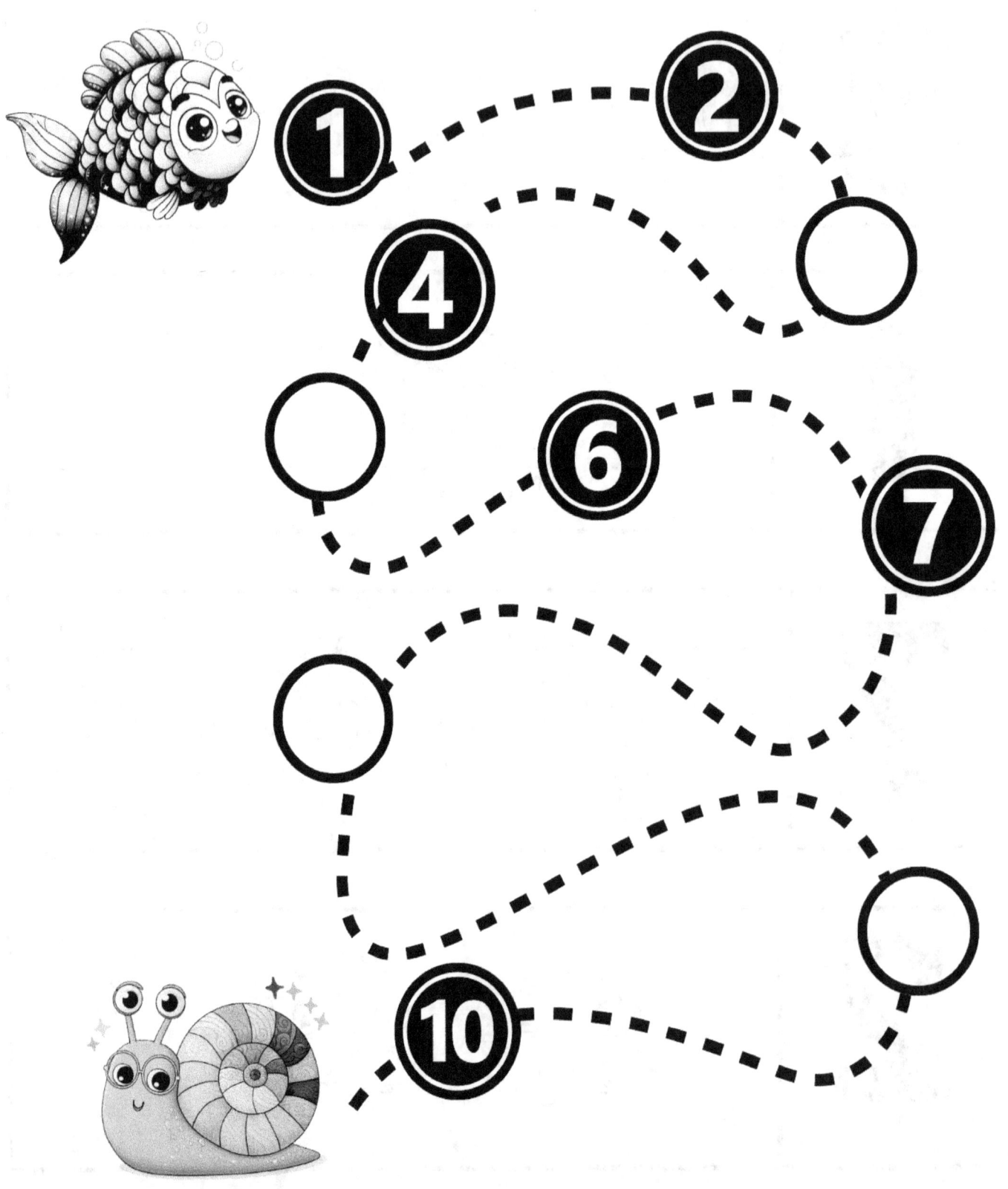

LET'S COUNT

Count the different objects and write your answers in the chart below

NUMBER AFTER

Fill in the number that comes after:

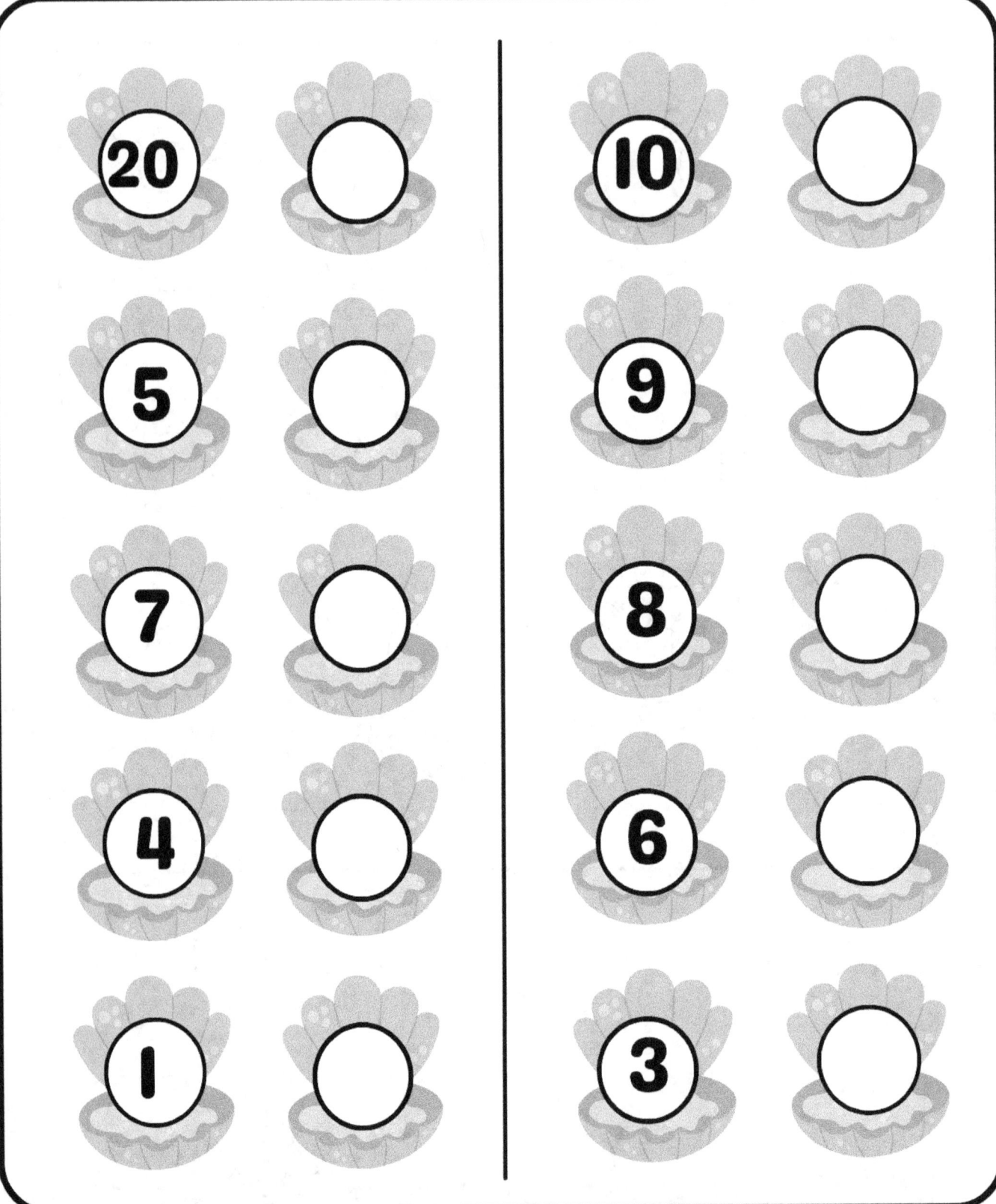

COUNT THE DOTS

Count the dots and write your answer.

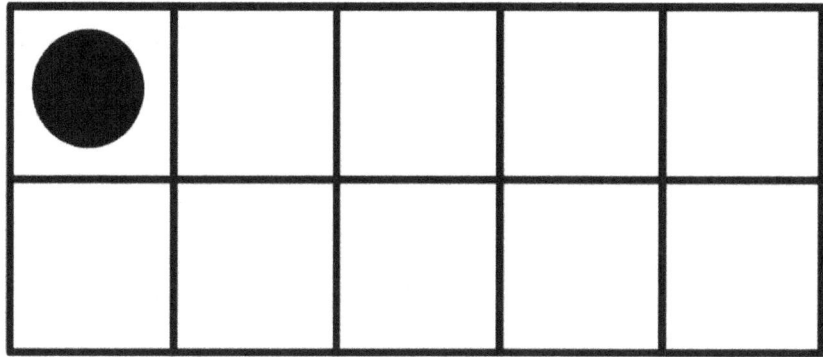

——

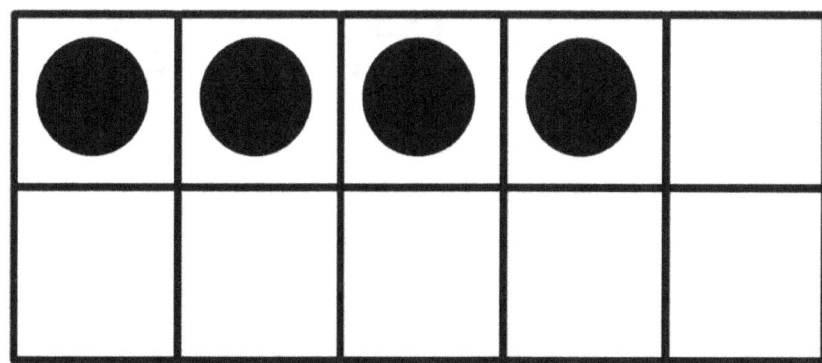

——

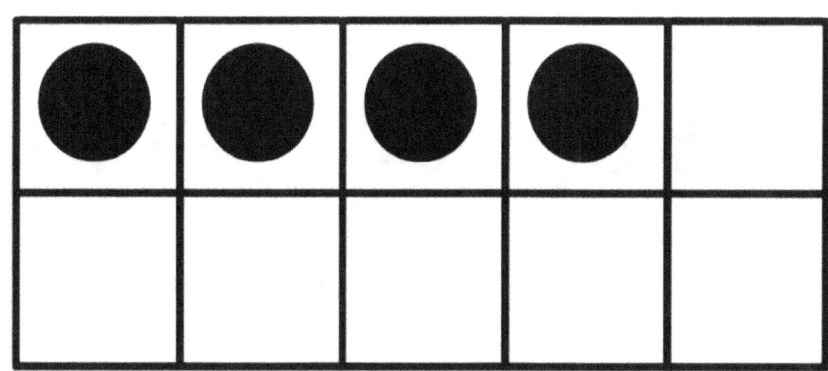

——

COUNT BY 2'S

Fill in the numbers to complete the sequence.

19   21   23   25   27

29   ☐   33   35   ☐

☐   ☐   ☐   45   47

49   51   53   55   57

COUNT AND COLOR

Count the images and color the correct answer

COUNT FORWARD

Fill in the numbers to complete the sequence.

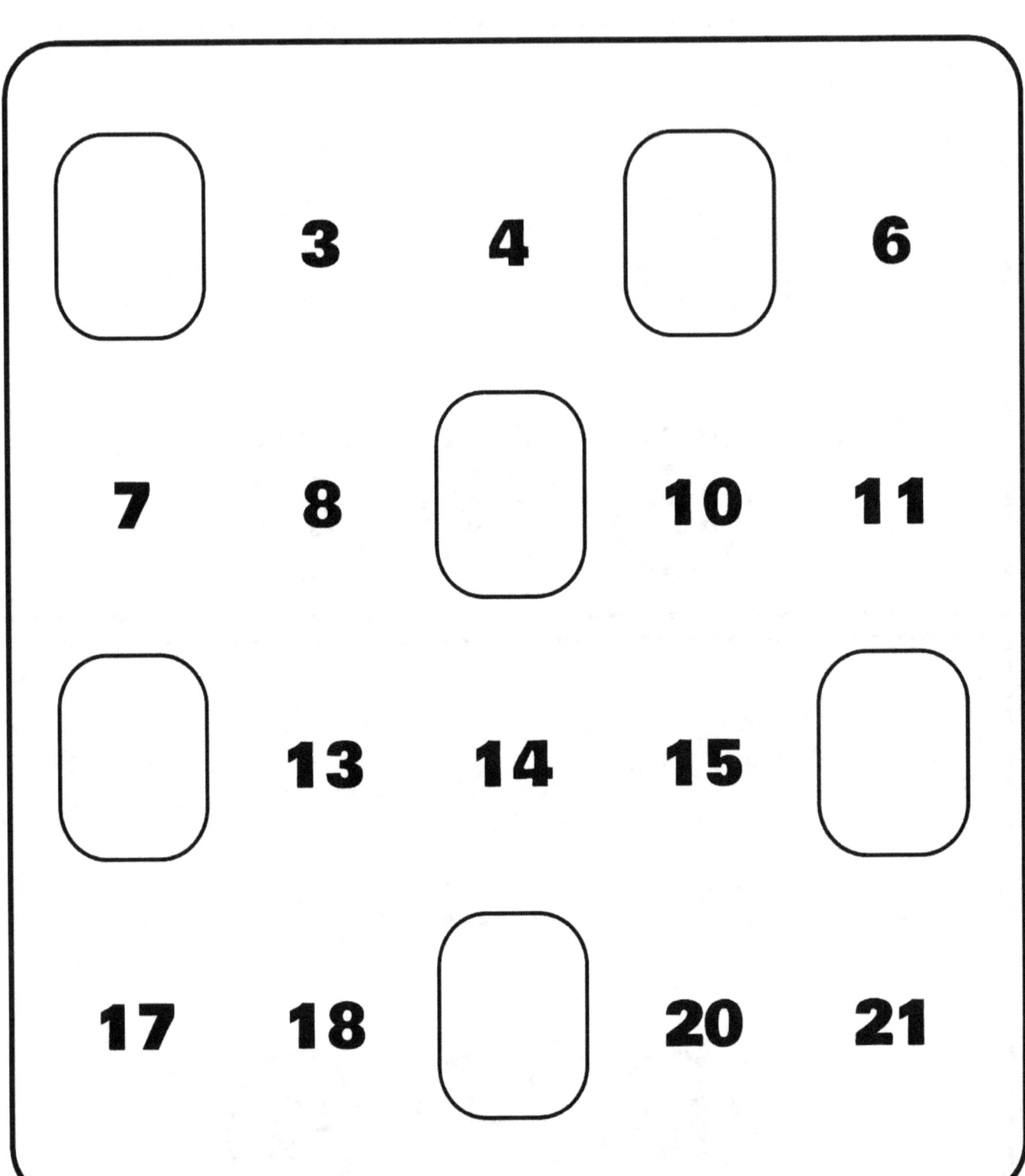

NUMBER RECOGNITION

Find the numbers and color in with the color given.

6 Purple **7** Light Blue **8** Gray **9** Pink **10** Blue

COUNT AND COLOR

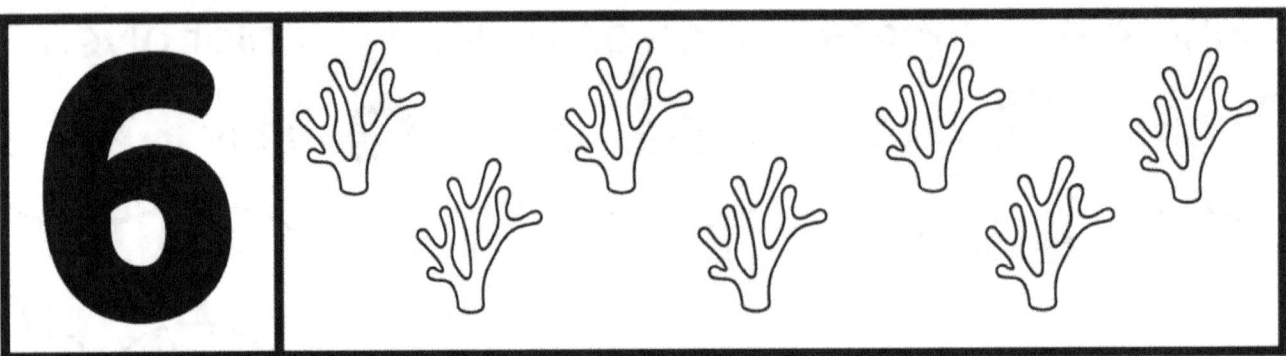

6

4

3

5

MISSING NUMBERS

Help Finley meet Ollie the Octopus by filling in the missing numbers on the path.

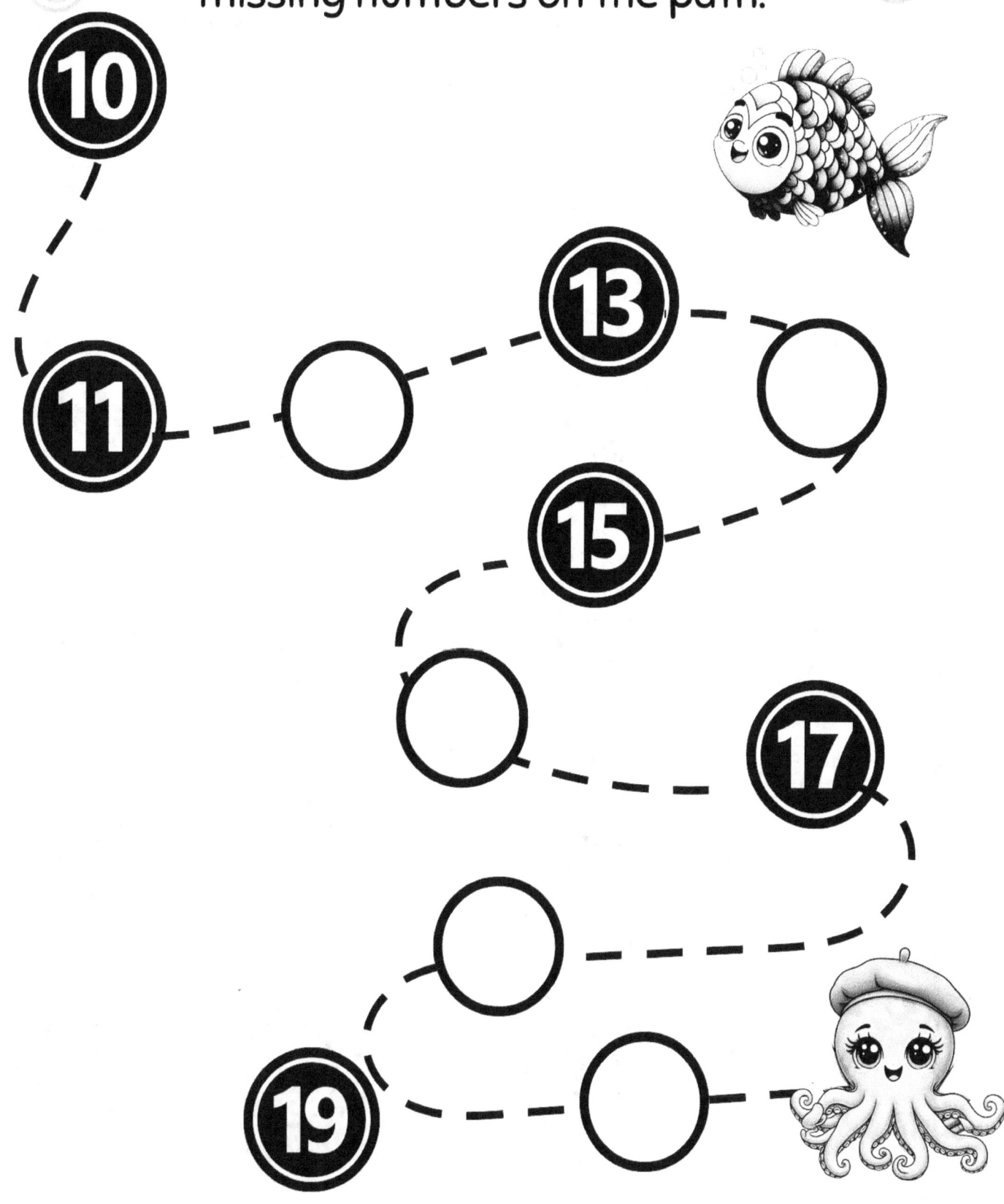

NUMBER BEFORE

Fill in the number that comes before:

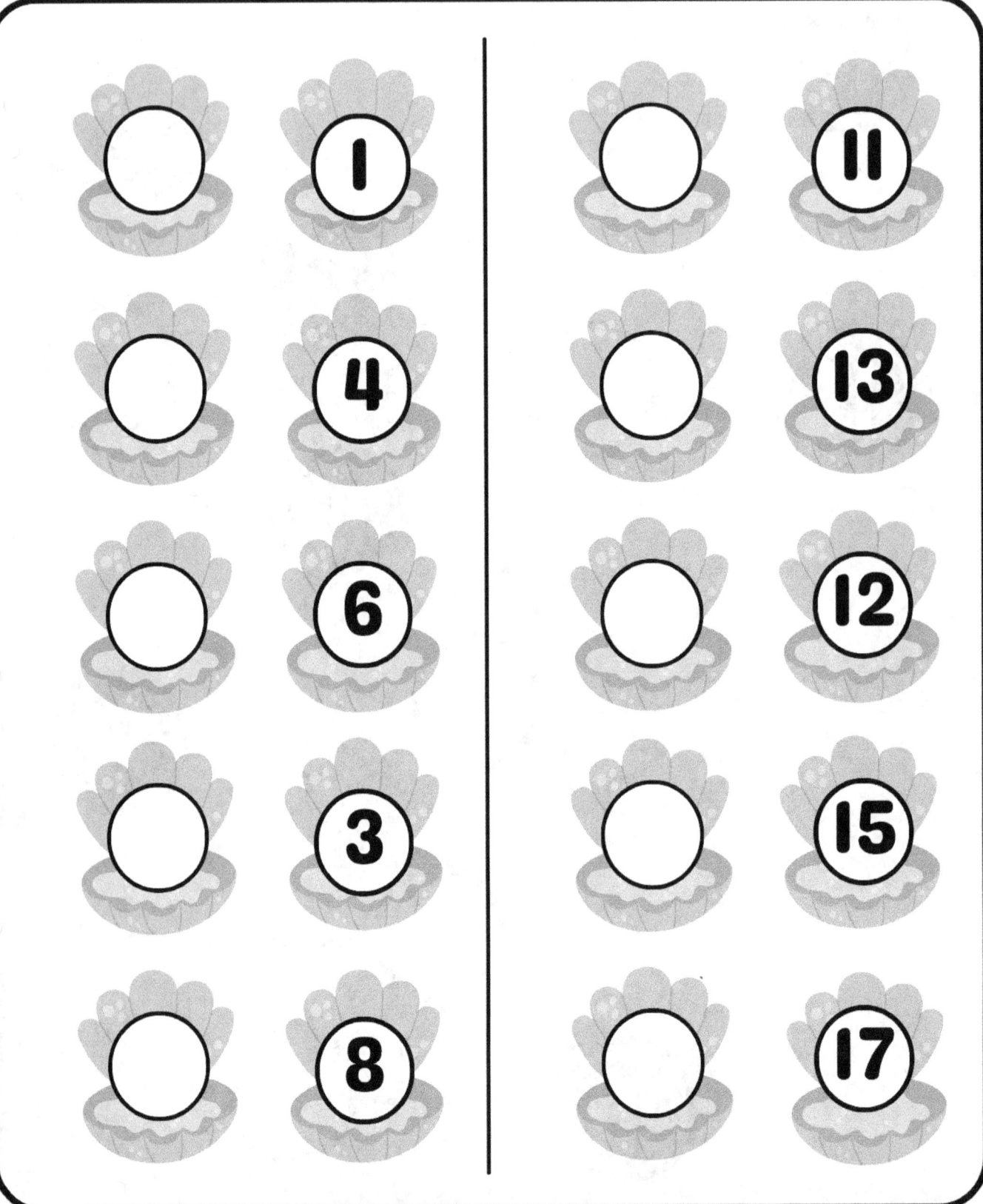

COUNT THE DOTS

Count the dots and write the answer

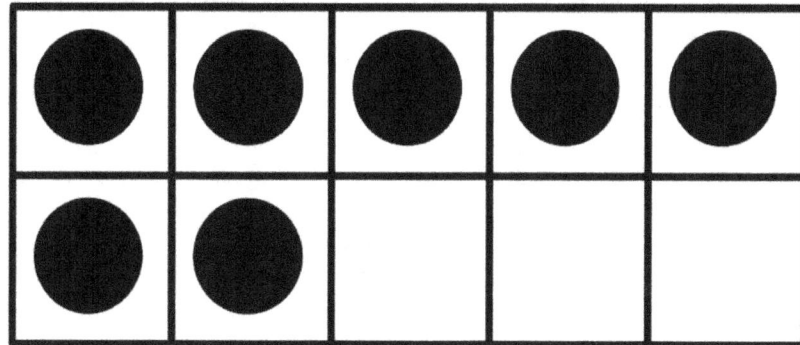

___

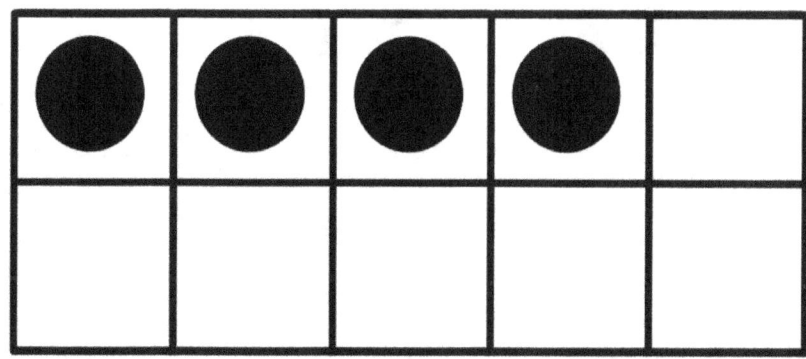

___

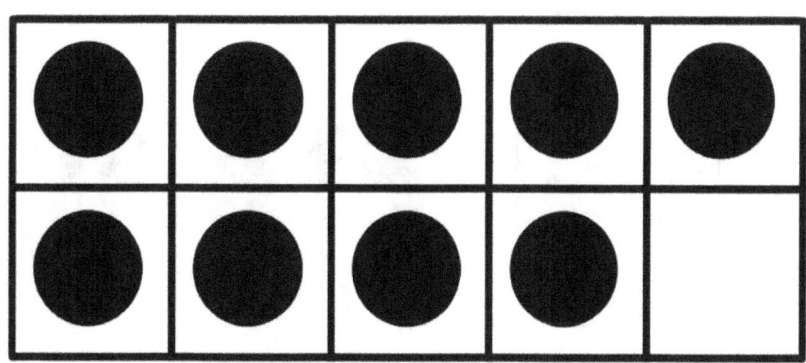

___

COUNT BY 2'S

Fill in the numbers to complete the sequence.

[ ] [ ] 9 11 13

15 17 19 [ ] 23

25 27 29 31 33

35 [ ] 39 41 [ ]

COLOR THE DOTS

Count and color the number of dots
for each number below.

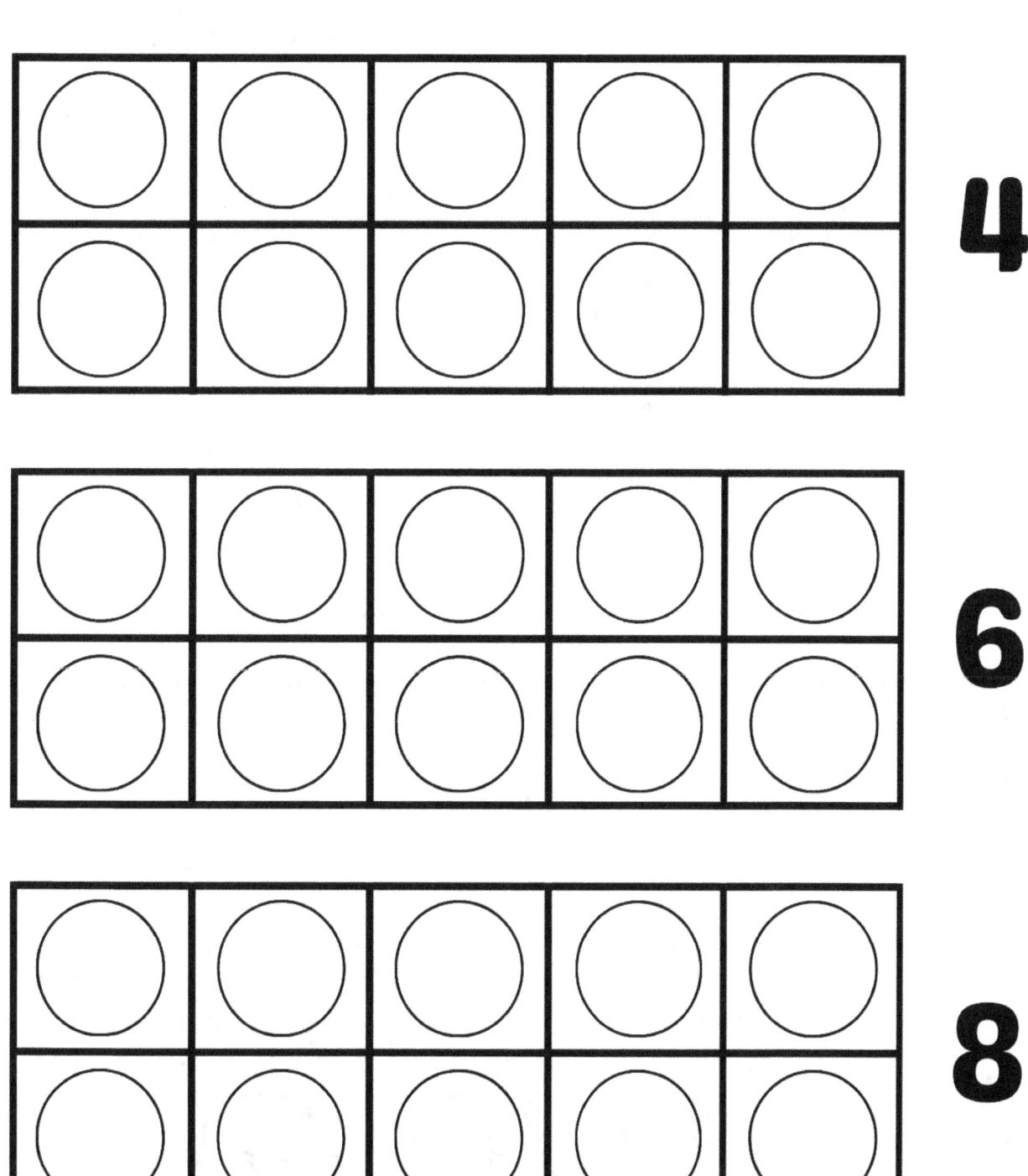

COUNT AND COLOR

Count the images and color the correct answer

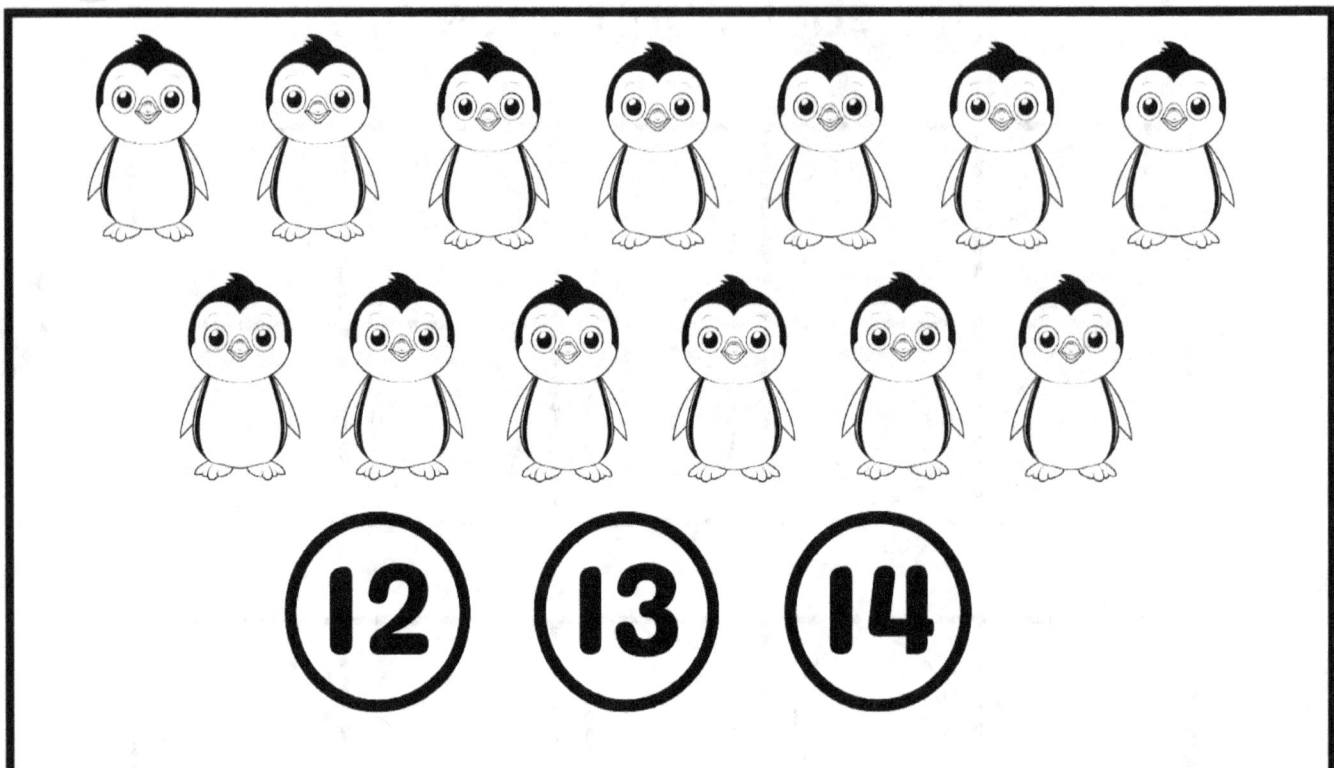

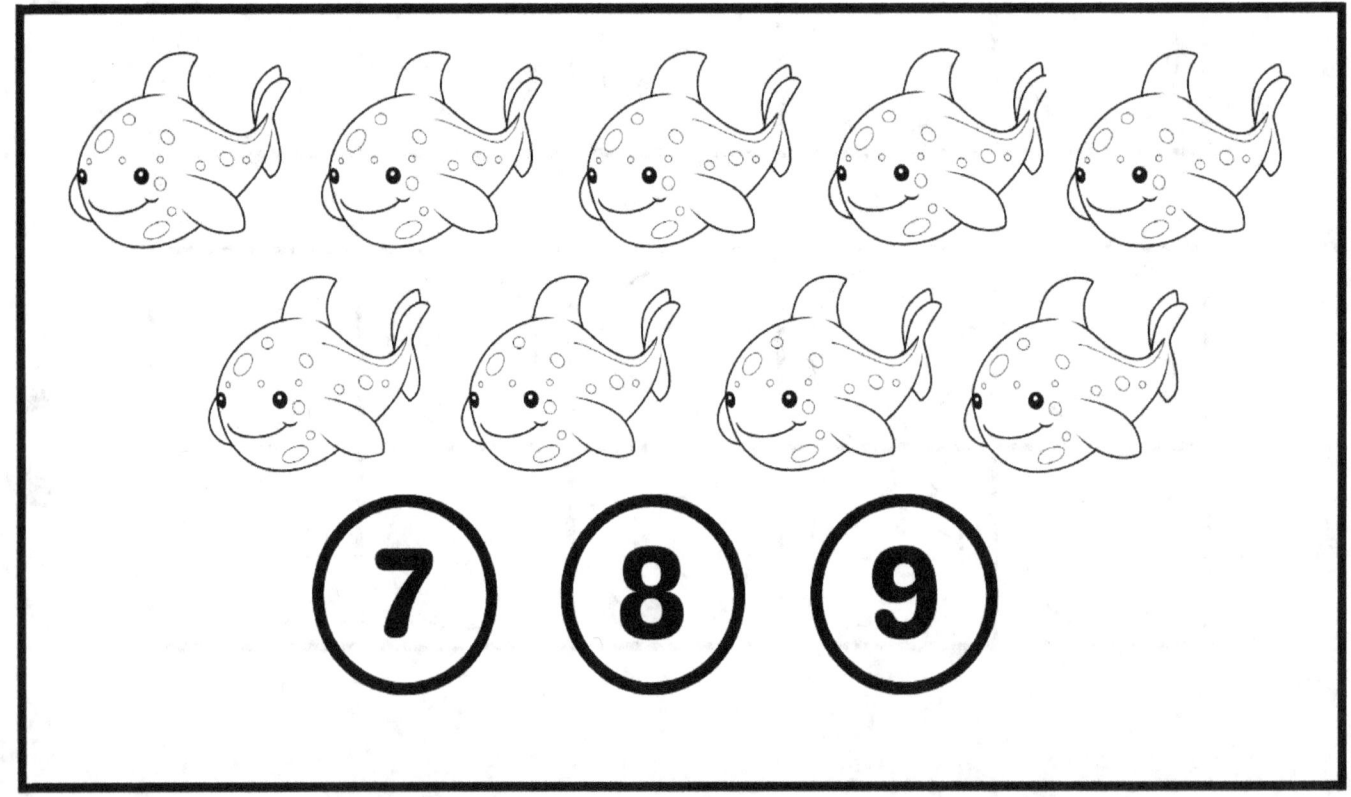

NUMBER RECOGNITION

Find the numbers and color in with the color given.

1 Green **2** Blue **3** Yellow **4** Purple **5** Light Blue

COUNT AND COLOR

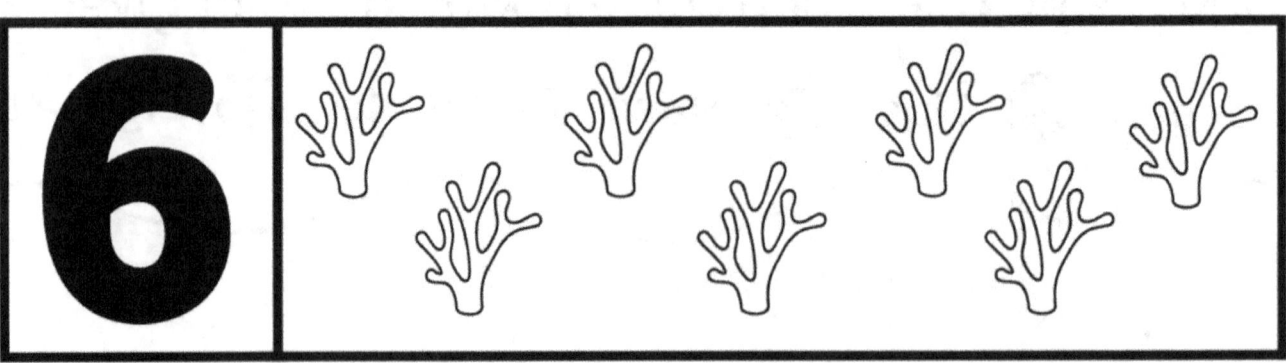

6

4

3

5

NUMBER ORDER

Fill in the missing number in the sequence.

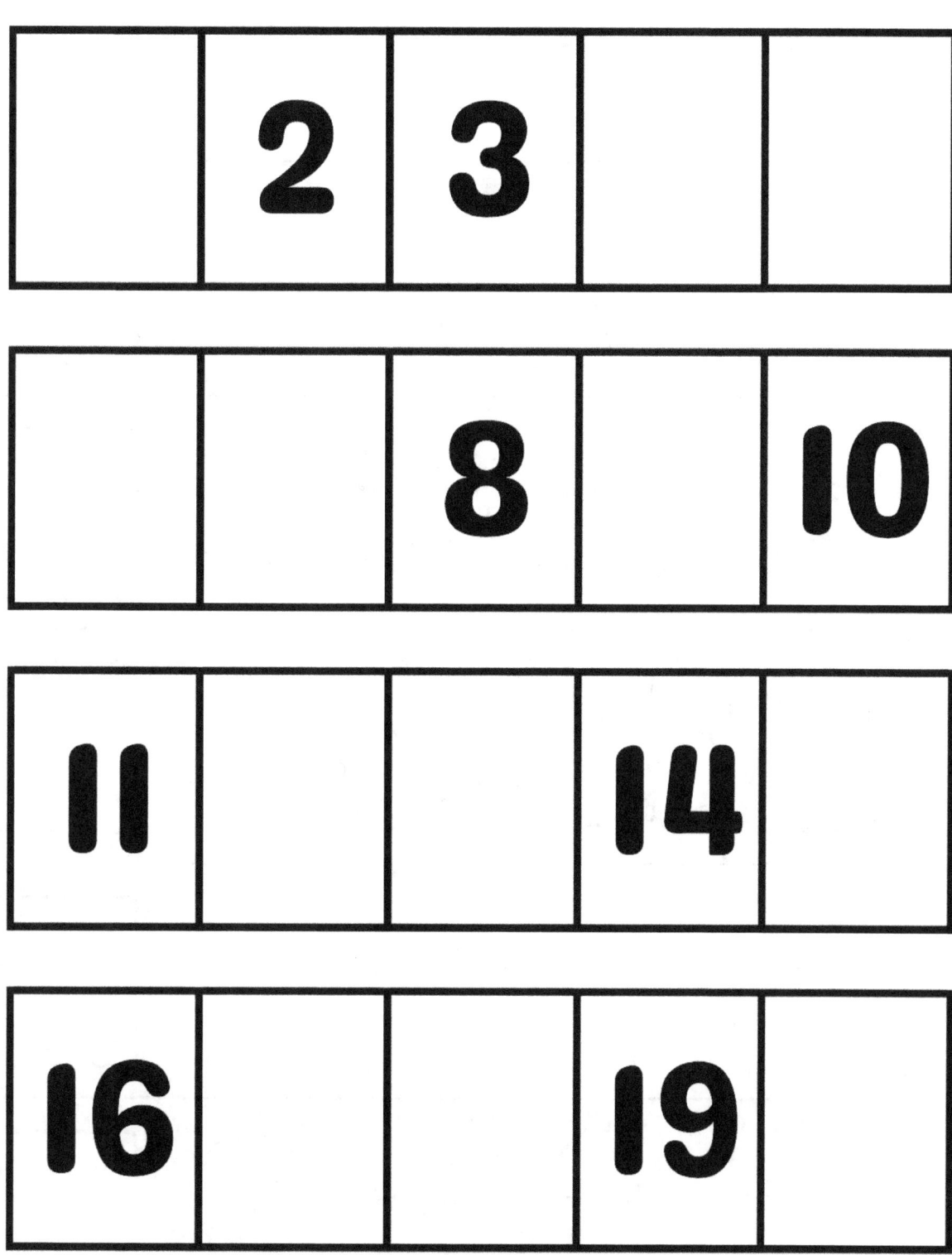

| | 2 | 3 | | |

| | | 8 | | 10 |

| 11 | | | 14 | |

| 16 | | | 19 | |

HOW MANY?

Count and write your answers in the box.

COUNT THE DOTS

Count the dots and write the answer

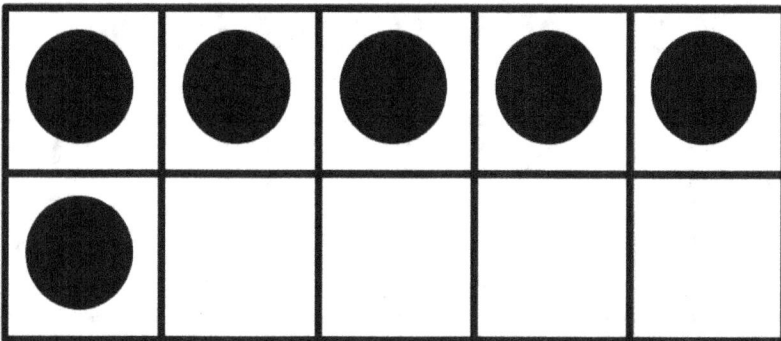

_____

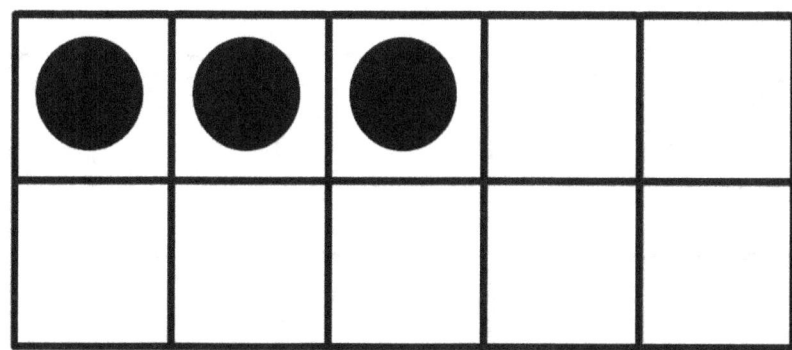

_____

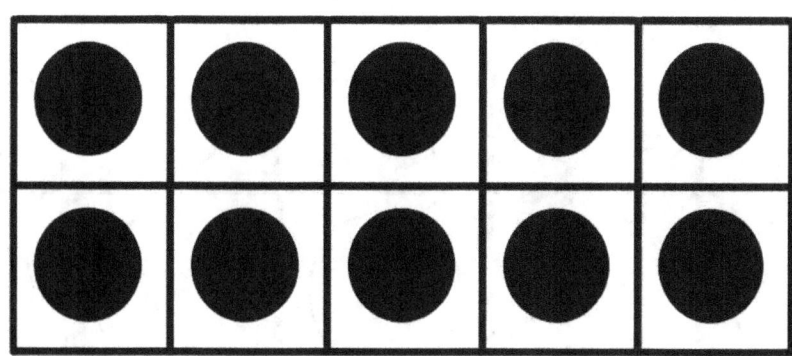

_____

COLOR THE DOTS

Count and color the number of dots
for each number below.

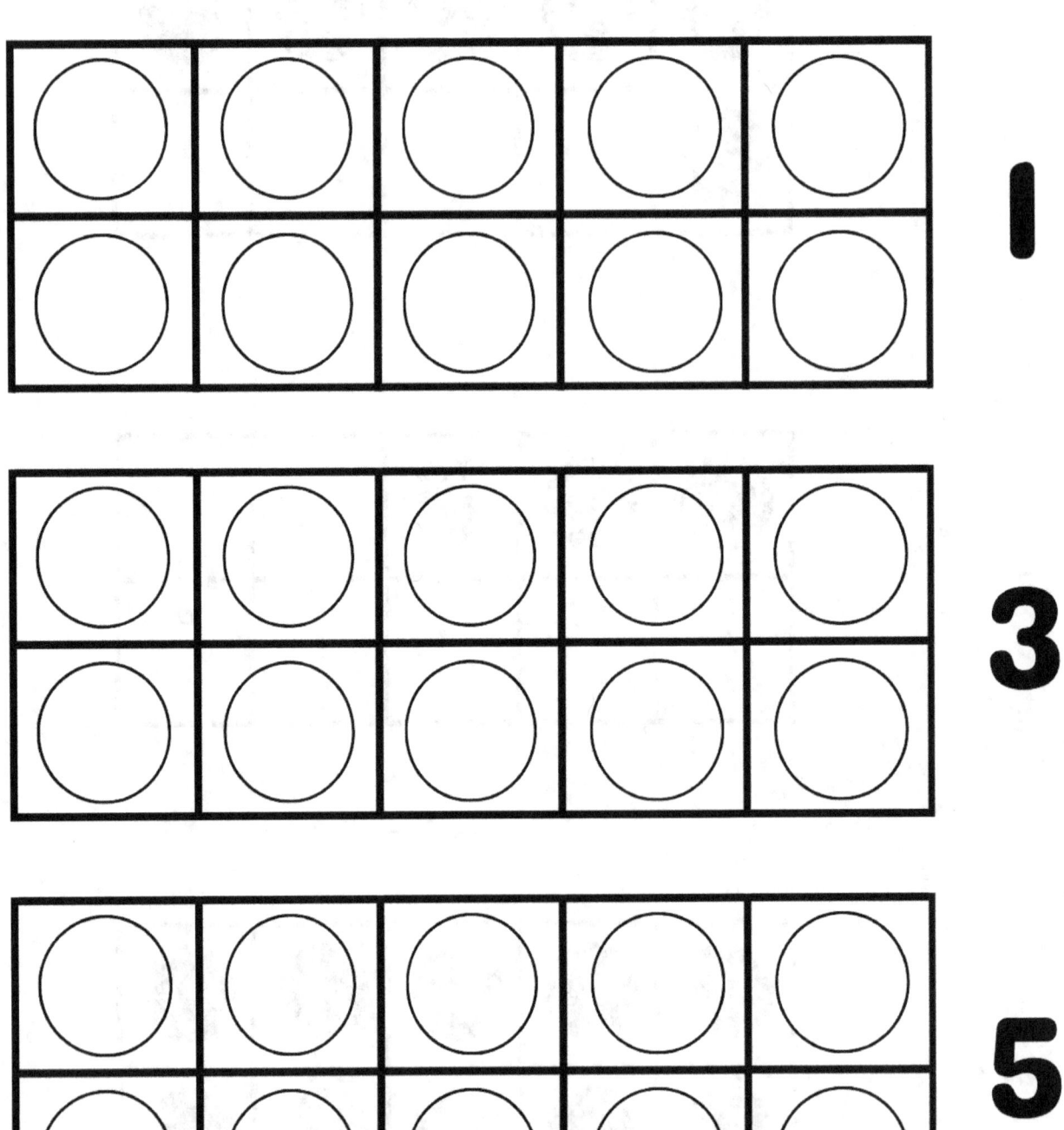

COUNT AND COLOR

Count the images and color the correct answer

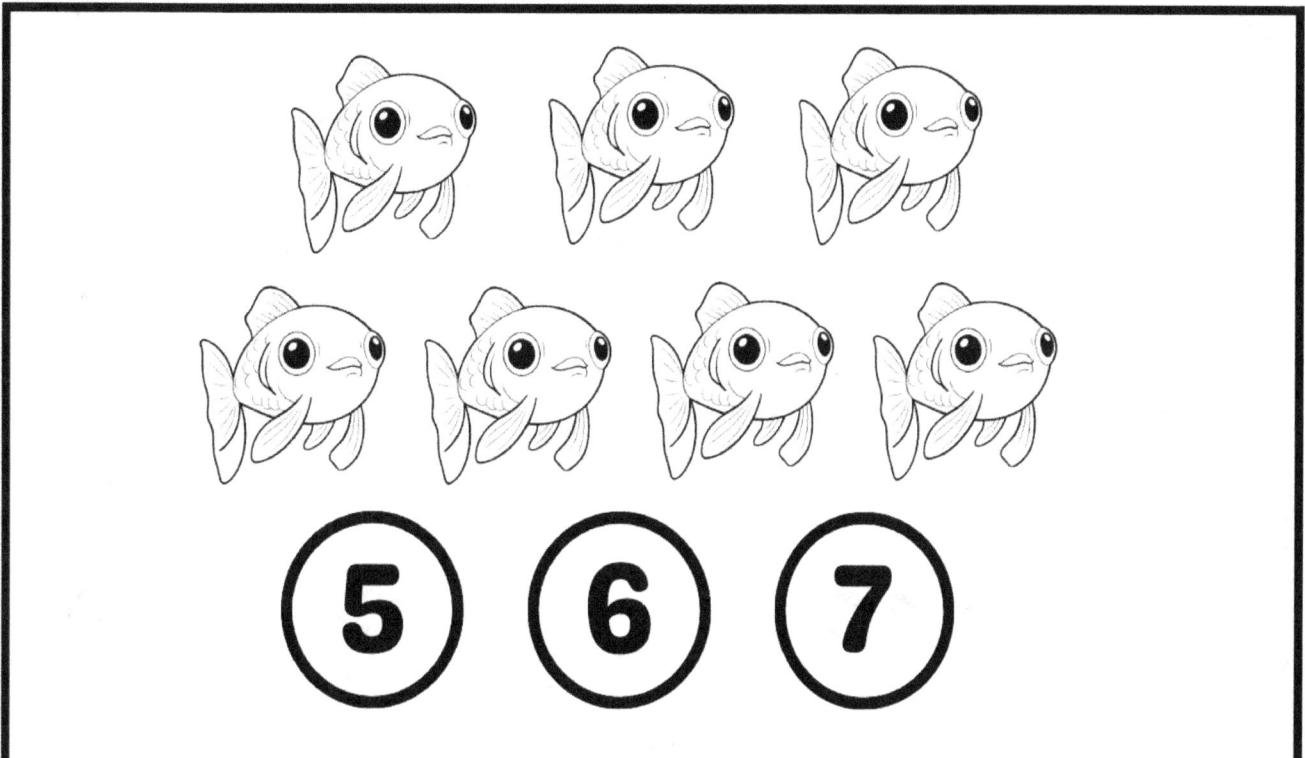

COUNT FORWARD

Fill in the numbers to complete the sequence.

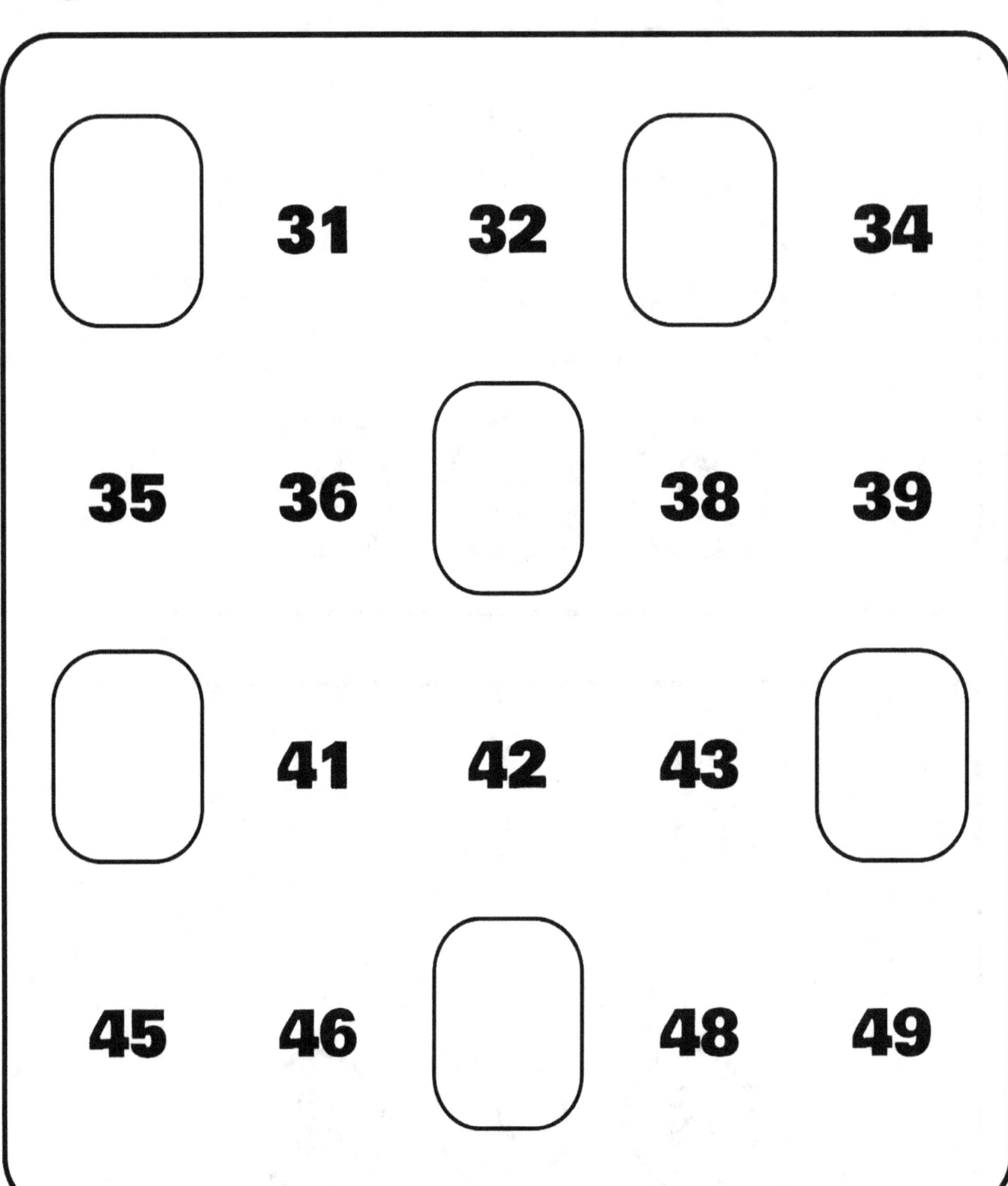

31 32 [ ] 34

35 36 [ ] 38 39

[ ] 41 42 43 [ ]

45 46 [ ] 48 49

MISSING NUMBERS

Fill in the missing numbers to help Finley find Harmony Hedgehog.

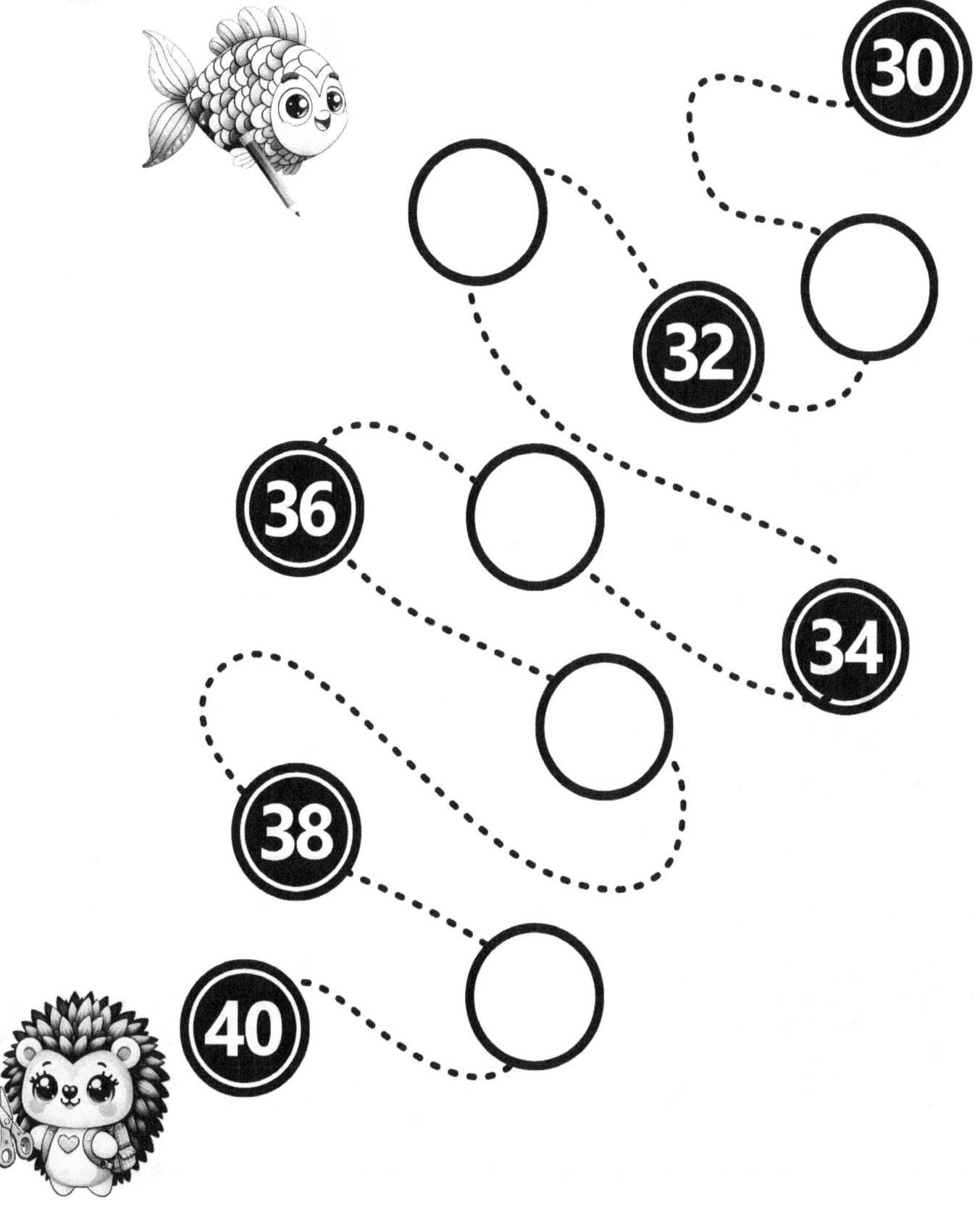

COUNT AND COLOR

15

13

16

14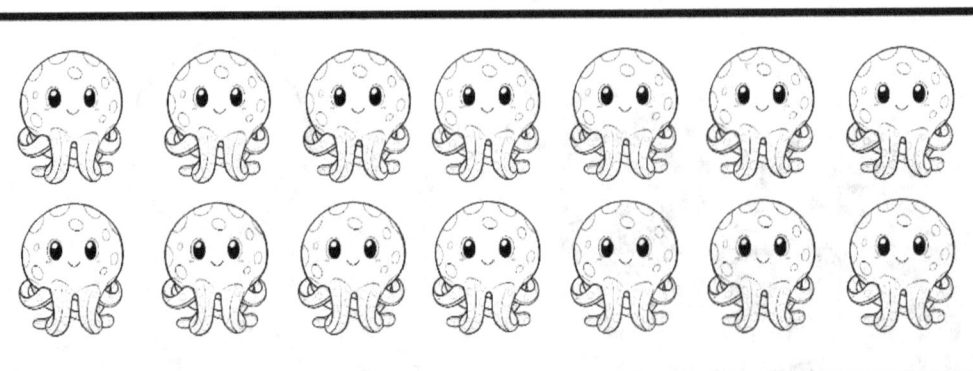

10 FRAME MATCH

Draw a line from the 10 frame to the matching number.

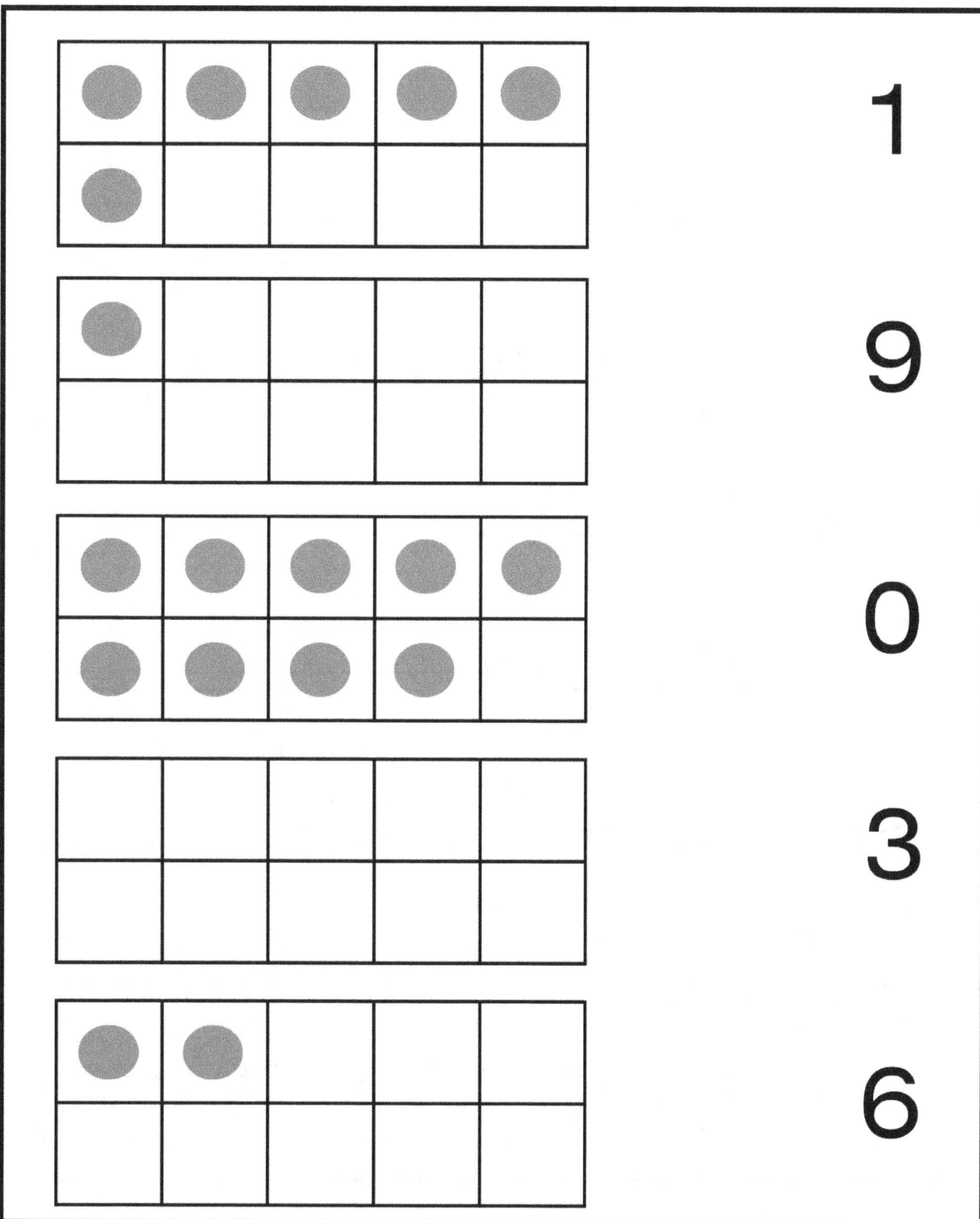

COUNT AND COLOR

Color the numbers of ocean shapes

5	
3	
7	
6	
9	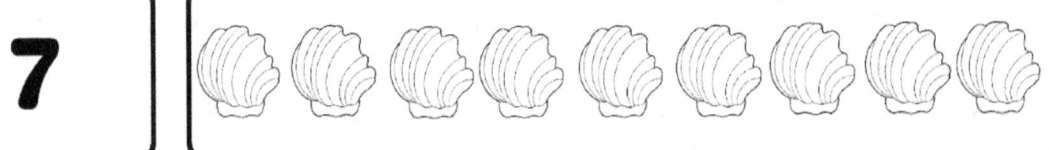
8	
4	

NUMBER AFTER

Fill in the number that comes after:

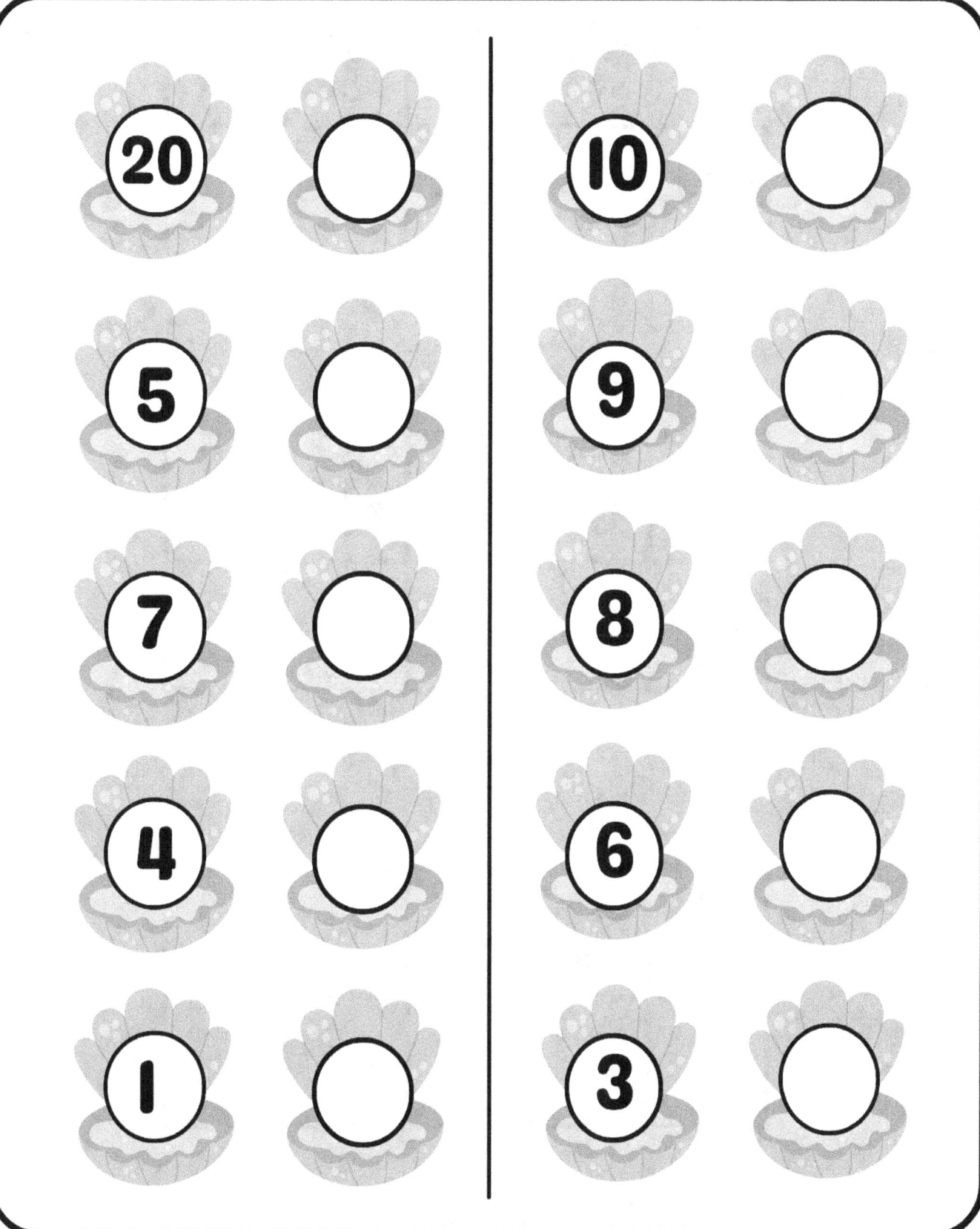

COUNT THE DOTS

Count the dots and write the answer

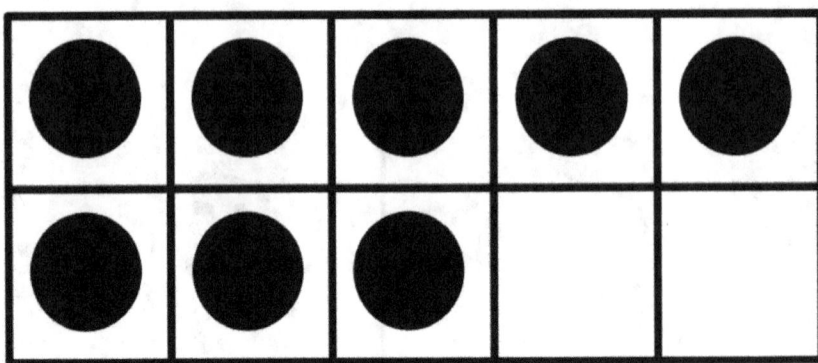

———

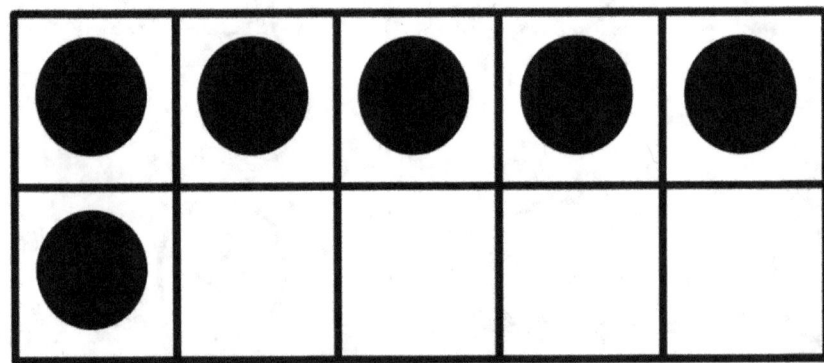

———

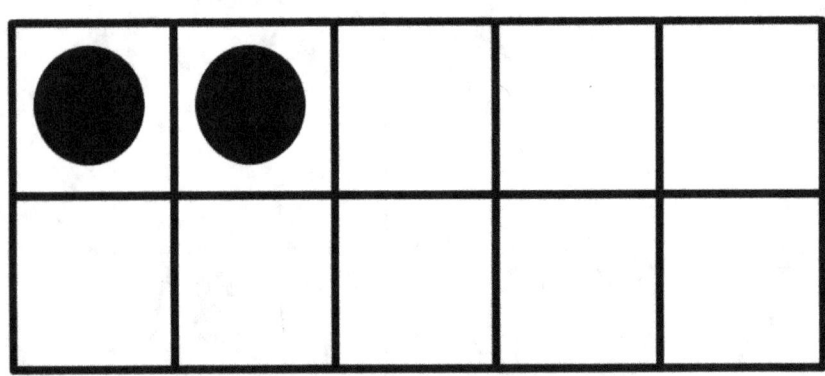

———

COUNT AND COLOR

Count the images and color the correct answer

COUNT FORWARD

Fill in the numbers to complete the sequence.

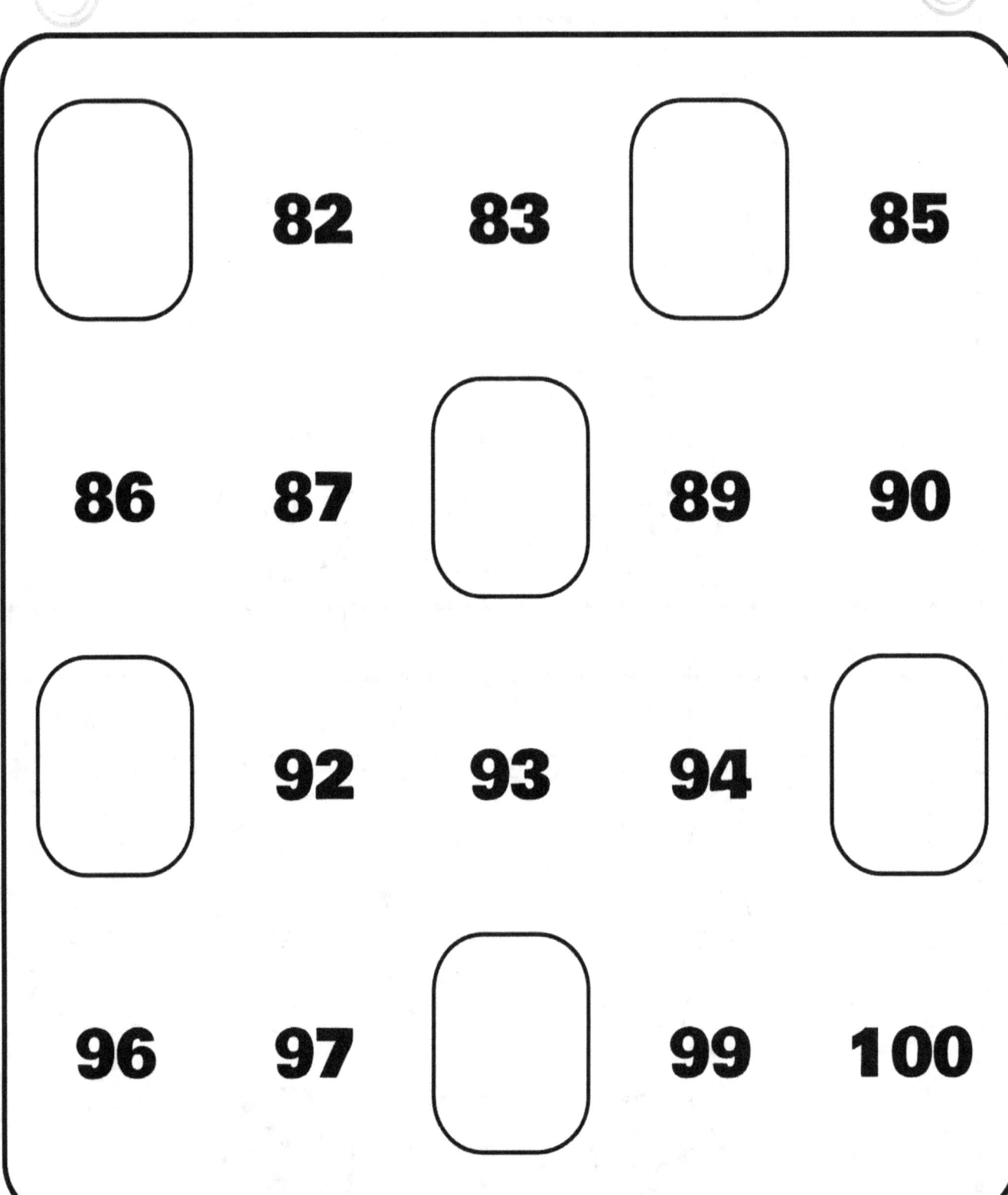

☐ 82 83 ☐ 85

86 87 ☐ 89 90

☐ 92 93 94 ☐

96 97 ☐ 99 100

HOW MANY FISH ?

Circle the correct number to match the number of fish inside the jar.

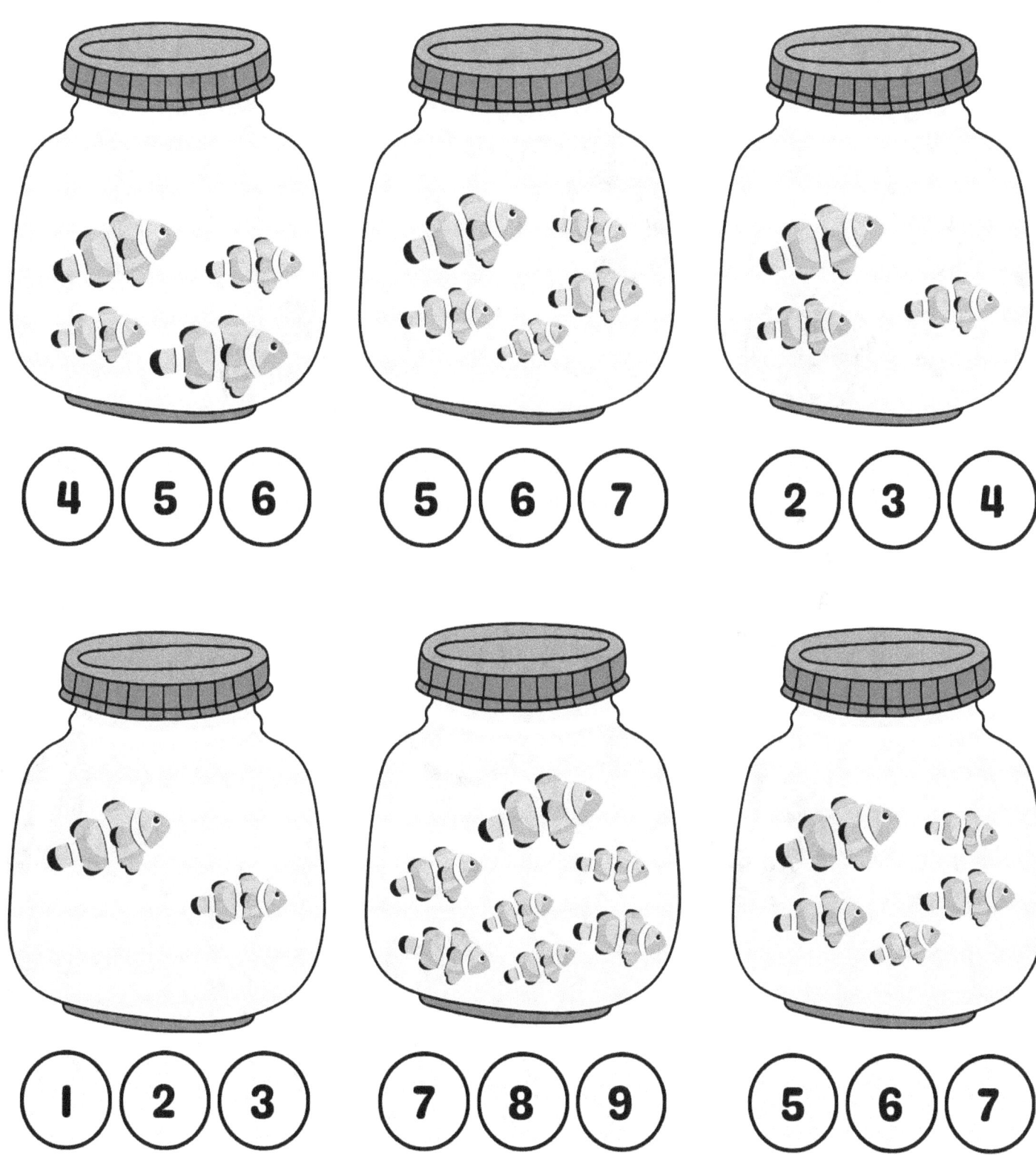

MISSING NUMBERS

Fill in the missing numbers to help Finley find Olivia Owlington.

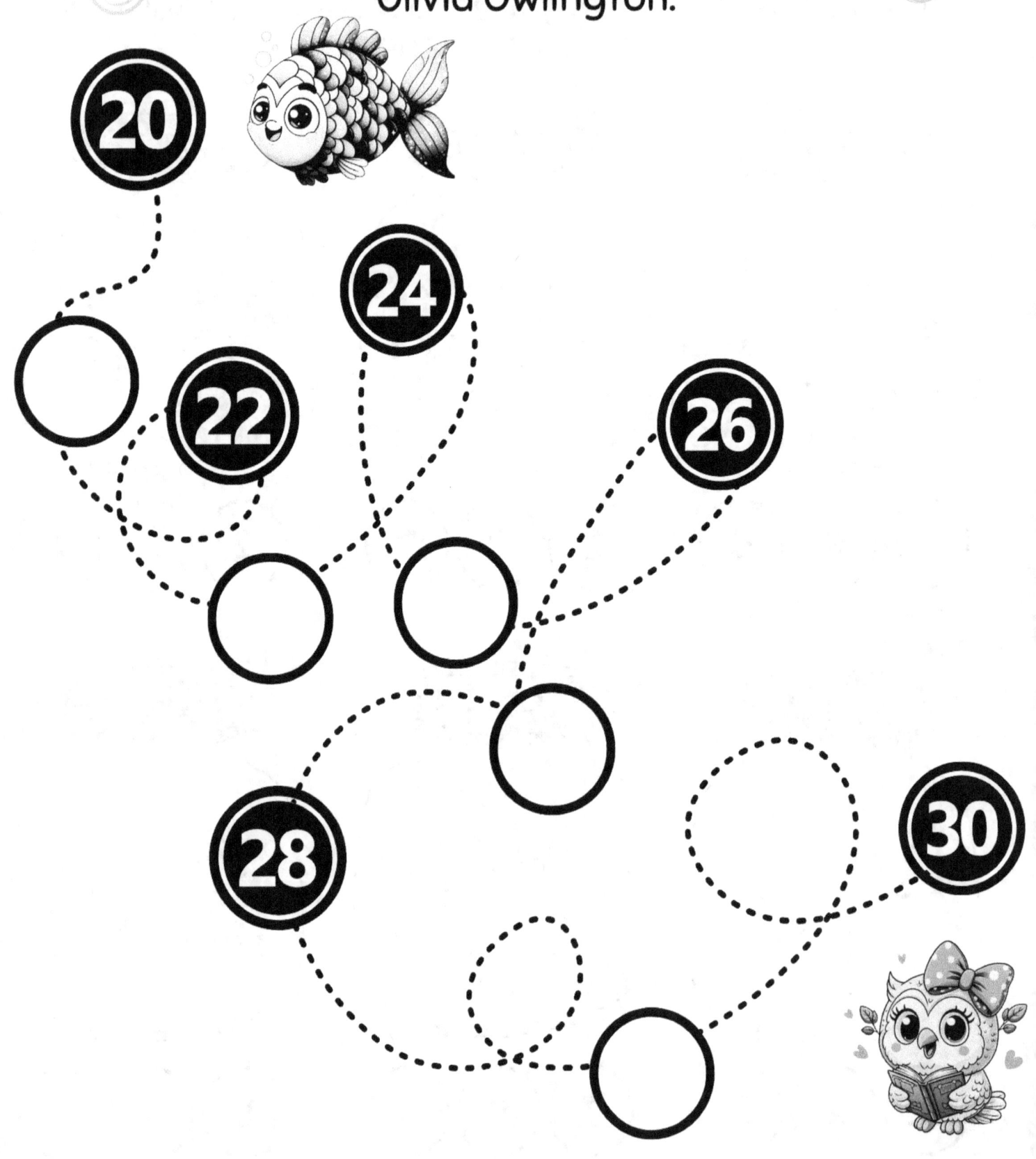

FIND AND COLOR

Find and color in all the fish with numbers
GREATER than 10.

LET'S COUNT
Color the circles that equal each number:

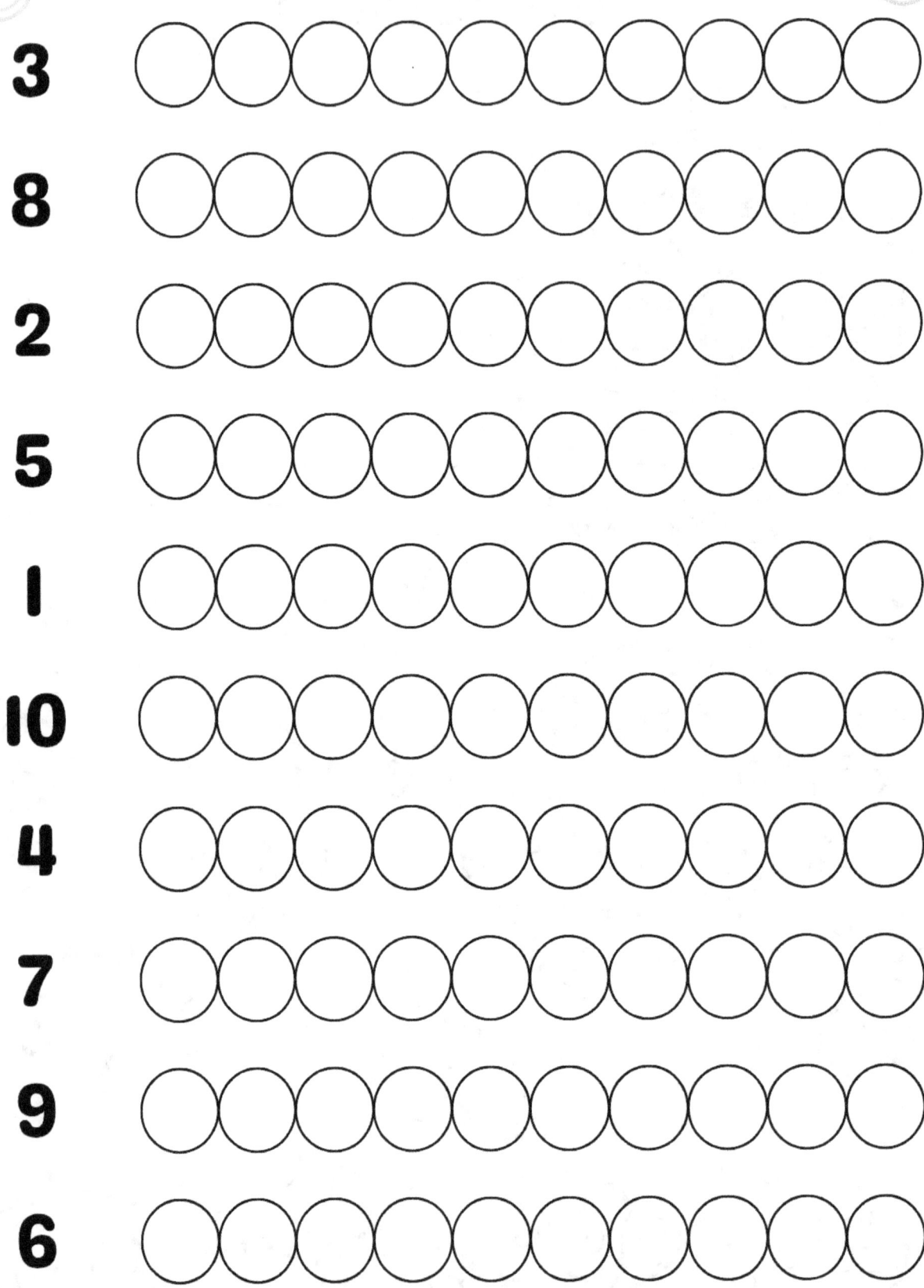

3

8

2

5

1

10

4

7

9

6

10 FRAME MATCH

Draw a line from the 10 frame to the matching number.

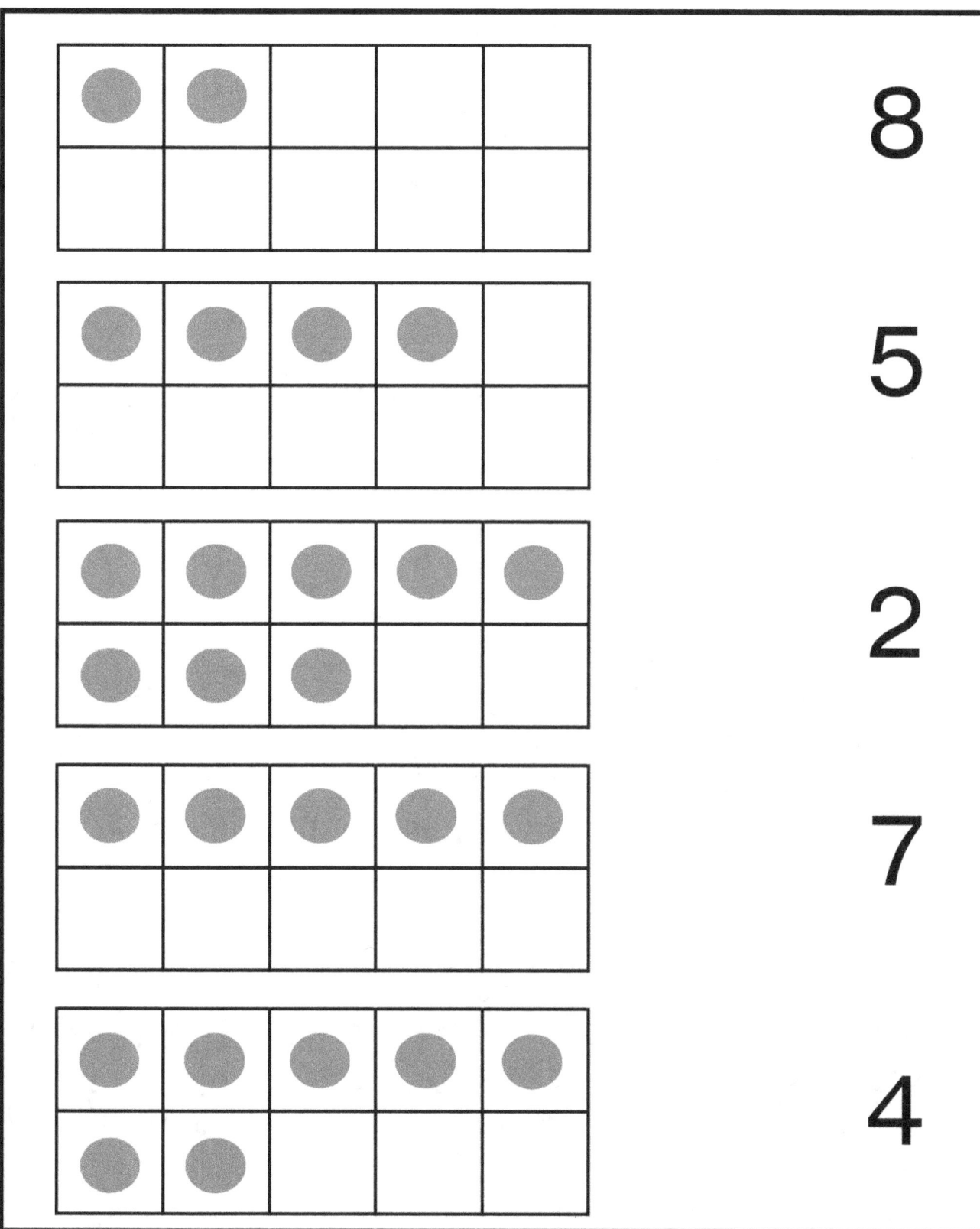

LET'S COUNT

Practice your counting and number writing by tracing the numbers in the fish below. When finished, color the fish.

COLOR ME !

COLOR ME !

COLOR ME !

COLOR ME !

www.ingramcontent.com/pod-product-compliance
Lightning Source LLC
Chambersburg PA
CBHW081103290526
45795CB00006B/1977